AAAS Handbook 1993/1994

Officers
Organization
Activities

American Association for the Advancement of Science

1333 H Street, NW, Washington, DC 20005

ISBN 0-87168-542-6
AAAS Publication 93-06S
Copyright 1993

American Association for the Advancement of Science
1333 H Street, N.W., Washington, DC 20005
202-326-6400

Contents

About AAAS 1

Principal Staff 2

Part I. Officers, Committees, Representatives

Board of Directors 5

Council 6

Committees and Representatives 10

    Standing Committees 10
        Council Affairs • Executive Committee of the Board
        of Directors • Investment and Finance • Nominations

    Committees of the Board of Directors 12
        Annual Meeting Site Committee • Audit Committee
        • Compensation Committee • Development Committee
        • Health of the Scientific Enterprise Committee
        • Long-Range Planning Committee • Prize Committee

    Award Committees 14
        AAAS–Philip Hauge Abelson Prize • AAAS–Newcomb
        Cleveland Prize • AAAS Award for Behavioral Science
        Research • AAAS–Hilliard Roderick Prize • AAAS Scientific
        Freedom and Responsibility Award • AAAS Award for
        Public Understanding of Science and Technology
        • AAAS–Westinghouse Science Journalism Awards
        • AAAS Mentor Award • AAAS International Scientific
        Cooperation Award

    Board-Appointed Committees 17
        AAAS-ABA National Conference of Lawyers and Scientists
        • Opportunities in Science • Public Understanding of
        Science and Technology • Science and International Security
        • Science, Engineering, and Public Policy • Scientific
        Freedom and Responsibility

Representatives to Other Organizations 22
   American Association for Accreditation of Laboratory
   Animal Care • AAAS Commissioners to the Commission
   on Professionals in Science and Technology
   • Gordon Research Conferences Board of Trustees
   • Interciencia Association • National Selection Committee,
   National Inventors Hall of Fame

Science Editorial Board 23

Sections: Officers and Committee Members 24
(Elected officers, members-at-large, Council delegates, representatives
of affiliates)
   Agriculture 24
   Anthropology 25
   Astronomy 27
   Atmospheric and Hydrospheric Sciences 27
   Biological Sciences 29
   Chemistry 32
   Dentistry 34
   Education 35
   Engineering 37
   General Interest in Science and Engineering 39
   Geology and Geography 42
   History and Philosophy of Science 44
   Industrial Science 45
   Information, Computing, and Communication 46
   Linguistics and Language Sciences 48
   Mathematics 48
   Medical Science 49
   Pharmaceutical Sciences 52
   Physics 53
   Psychology 54
   Social, Economic, and Political Sciences 56
   Societal Impacts of Science and Engineering 58
   Statistics 61

Electorate Nominating Committees *(by Section)* 62

Regional Division Officers 69

Part II. Membership, Organization, Activities

Membership 72
      Fellows 72

Organization 74
      Sections 74
      Electorates 74
      Membership by Electorate 75
      Regional Divisions 75
          Arctic • Caribbean • Pacific
          • Southwestern and Rocky Mountain
      Affiliated Organizations 77
      Affiliated Academies of Science 81
      National Association of Academies of Science 82
      Participating Organizations 83
          Commission on Professionals in Science and Technology
          • Gordon Research Conferences • Illinois Science Lecture
          Association

Activities 85
      Publications 85
          Science • *The Online Journal of Current Clinical Trials*
          • *Science Books & Films* • Other Publications

      Meetings 87
          Annual National Meeting • Science Innovation • The Human
          Genome Conference Series • Topical Conferences
          • Colloquium on Science and Technology Policy • AAAS
          Forum for School Science • Science and Security Colloquium
          Regional Division Meetings

      Directorates 89
          Directorate for Education and Human Resources Programs 90
          Policy Development • Science, Mathematics, and
          Technology Education Programs • Linkages Programs • Public
          Understanding of Science and Technology

          Directorate for International Programs 95
          Global Change • Sub-Saharan Africa Program • Western
          Hemisphere Cooperation • International Scientific
          Cooperation • Program on Science and International
          Security (PSIS)

Project 2061 97

Directorate for Science and Policy Programs 97
Science, Technology, and Government • Science and
Human Rights • Scientific Freedom, Responsibility, and Law
• Science and Engineering Fellowship Programs

Awards, Prizes, and Grants 99
AAAS–Philip Hauge Abelson Prize • AAAS–Newcomb
Cleveland Prize • AAAS Award for Behavioral Science
Research • AAAS–Hilliard Roderick Award • AAAS Scientific
Freedom and Responsibility Award • AAAS Award for Public
Understanding of Science and Technology
• AAAS–Westinghouse Science Journalism Awards
• AAAS Mentor Award • AAAS International Scientific
Cooperation Award • AAAS Academy Research Grants
• AAAS William D. Carey Annual Science Award

Part III. History, Policy, Governance

History 111
    Founding 111
    Milestones of the AAAS 113
    Resolutions and Policy Statements 118
    Meetings and Presidents 121
    Administrative Officers, 1848-1993 126
    Former Members of the Board of Directors, 1964-1993 128

Policy Statements 130
    Equal Opportunity in the Sciences and Engineering 130
    Editorial Policy, *Science* Magazine 130
    Criteria for Affiliation with AAAS 131
    Procedures for Affiliation with AAAS 131
    Policy Concerning Council Meeting Agenda 132
    Summary of Policy, Guidelines, and Procedures for
    Communication with Congress 133

Governance 135
    Constitution 135
        Name • Objectives • Membership and Affiliation
        • Electorates • Sections • Officers • Council
        • Board of Directors • Amendments • Tax-Exempt Status

Bylaws 139
　　　　　Membership and Affiliation • Electorates • Sections
　　　　　• Officers • Nominations and Elections • Council
　　　　　• Board of Directors • Financial Administration
　　　　　• Publications • Scientific Meetings • Committees
　　　　　• Regional Divisions and Local Branches
　　　　　• Participating Organizations • Official Statements
　　　　　• Parliamentary Authority • Amendments

Articles of Incorporation 151
Index 153

About AAAS

Since its founding in 1848, the American Association for the Advancement of Science has continually worked to advance science. From its early, specific aims concerned with communication and cooperation among scientists, the Association's goals now encompass the broader purposes of ". . . furthering the work of scientists, facilitating cooperation among them, fostering scientific freedom and responsibility, improving the effectiveness of science in the promotion of human welfare, advancing education in science, and increasing the public understanding and appreciation of the importance of the methods of science in human progress."

Beginning with 461 founding members in 1848, AAAS now enrolls over 134,000 scientists, engineers, science educators, policymakers, and others interested in science and technology who live and work in the United States and in many other countries throughout the world. In addition, AAAS is the world's largest federation of scientific and engineering societies, with 292 organizations that cooperate with the Association on a variety of projects, including Annual Meeting symposia, fellowships, international programs, annual analyses of the federal research and development budget, equal opportunity activities, and science education.

AAAS is directly governed by an elected Board of Directors and its general policies are set by a Council representing the 23 AAAS sections, regional divisions, and science academies. The AAAS section structure allows members to align themselves with specific disciplinary areas within the Association, ranging from the physical, biological, and health sciences to the social, economic, and applied sciences and mathematics. Through the sections, members work on both disciplinary and interdisciplinary concerns. The four geographic divisions of AAAS Pacific, Southwestern and Rocky Mountain, Arctic, and Caribbean offer additional opportunities for member involvement.

A central staff of some 270 people, headquartered in Washington, DC, handles the Association's day-to-day activities, including the editing and production of *Science* and other publications; planning and support of the Annual Meeting and a variety of colloquia and other meetings; development of special programs in science education and human resources, international scientific cooperation, and science and public policy; management of a variety of fellowships, grants, and prizes; and dissemination of information about all of these activities.

Principal staff and phone numbers are listed on page 2.

Principal Staff

EXECUTIVE OFFICE: 202-326-6640; FAX: 202-371-9526
    Executive Officer and Publisher: Richard Nicholson
    Chief Financial & Administrative Officer: Carl Amthor
    Science Advisor: Philip Abelson
    Education Advisor: James Rutherford
    Executive Assistant to the Executive Officer: Gretchen Seiler

DIRECTORATES

Education and Human Resource Programs: 202-326-6670; FAX: 202-371-9849
    Director: Shirley Malcom
International Programs: 202-326-6650; FAX: 202-289-4958
    Director: Richard Getzinger
Project 2061: 202-326-6666; FAX: 202-842-5196
    Director: James Rutherford
Science and Policy Programs: 202-326-6600; FAX: 202-289-4950
    Director: Albert Teich

OFFICES

Communications: 202-326-6440; FAX: 202-789-0455
    Acting Director: Nan Broadbent
Development: 202-326-6636; FAX: 202-371-9849
    Director: Jeannette Wedel
Financial Services: 202-326-6699; FAX: 202-842-1711
    Director: Robert Vaughan
Administration: 202-326-6470; FAX: 202-682-1630
    Director: Debra Wright
Meetings: 202-326-6450; FAX: 202-289-4021
    Director: Robin Woo
Membership and Circulation: 202-326-6430; FAX: 202-842-1065
    Director: Michael Spinella
Publications: 202-326-6460; FAX: 202-842-1065
    Director: Patricia Morgan
***Science* Business Office:** 202-326-6540; FAX: 202-682-0816
    Associate Publisher: Beth Rosner
***Science* Editorial Center:** 202-326-6500; FAX: 202-289-7562 or 202-371-9227
    Editor-in-Chief: Daniel Koshland, Jr.
    Editor: Ellis Rubinstein
    Managing Editor: Monica Bradford
The AAAS main number is 202-326-6400.

I
Officers
Committees
Representatives

Board of Directors

The Board of Directors is the legal representative of the Association and is responsible for conducting its affairs. It consists of 13 members: the chairman of the Board (the retiring president of AAAS), the president, the president-elect, the treasurer, eight other directors, and the executive officer (without vote). Annually, the membership elects a president-elect, who serves on the Board for three years, and two other new members for four-year terms. The executive officer and the treasurer are appointed by and serve at the pleasure of the Board, which meets five times a year. The Executive Committee (see page 10) acts on behalf of the Board at other times. The executive officer and staff manage the affairs of the Association in accordance with procedures determined by the Board.

Officers

Retiring President and Chairman: F. Sherwood Rowland (1994), Dept. of Chemistry, University of California, Irvine, CA 92717
President: Eloise E. Clark (1995), Academic Affairs, 230 McFall Center, Bowling Green State University, Bowling Green, OH 43403
President-Elect: Francisco J. Ayala (1996), Dept. of Ecology and Evolutionary Biology, Rm. 716 Engineering Bldg., University of California, Irvine, CA 92717
Treasurer: William T. Golden, 40 Wall St., Rm. 4201, New York, NY 10005
Executive Officer: Richard S. Nicholson, AAAS (ex officio)

Other Members

Robert A. Frosch (1994), General Motors Research, General Motors Corp., Warren, MI 48090-9055
Florence P. Haseltine (1994), National Institutes of Health–NICHD, 6100 Executive Blvd., Rm. 8B07, Rockville, MD 20852
William A. Lester, Jr. (1997), Dept. of Chemistry, University of California, Berkeley, CA 94720
Alan Schriesheim (1996), Argonne National Laboratory, 9700 South Cass Ave., Bldg. 201, Argonne, IL 60439-4832
Jean'ne M. Shreeve (1995), University Research Office, 111 Morrill Hall, University of Idaho, Moscow, ID 83844-3010
Chang-Lin Tien (1996), Office of the Chancellor, 200 California Hall, University of California, Berkeley, CA 94720
Warren M. Washington (1995), National Center for Atmospheric Research, P.O. Box 3000, Boulder, CO 80307-3000
Nancy S. Wexler (1997), Depts. of Neurology and Psychiatry, College of Physicians and Surgeons of Columbia University, 722 West 168th Street, New York, NY 10032

Note: Terms end on the last day of the Annual Meeting held in the years given in parentheses.

Council

The Council is responsible for establishing the general policies governing all programs of the Association. All members of the Board of Directors (see page 5) are also members of the Council, as are the retiring section chairs, who serve as representatives of their section committees. In addition, the Council has one or more delegates from each electorate; one delegate from each regional division; and two delegates from the National Association of Academies of Science a total of 81 persons. The AAAS president serves as chair and the executive officer as secretary. Delegates may serve a maximum of two consecutive three-year terms. The Council meets during the Annual Meeting of the Association. The Committee on Council Affairs (see page 10) serves as the executive committee of the Council.

Council Chair

Eloise E. Clark, 230 McFall Center, Bowling Green State University, Bowling Green, OH 43403

Secretary to the Council

Richard S. Nicholson, AAAS executive officer

Members of the Board of Directors are also members of the Council.

Other Council Members

George J. Armelagos (1995; *Electorate on Anthropology*), Dept. of Anthropology, Emory University, Atlanta, GA 30322

Walter S. Baer (1994; *Section on Industrial Science*), The RAND Corp., 1700 Main St., P.O. Box 2138, Santa Monica, CA 90406-2138

Victor R. Baker (1994; *Section on Geology and Geography*), Dept. of Geosciences, University of Arizona, Tucson, AZ 85721

Allen J. Bard (1995; *Electorate on Chemistry*), Dept. of Chemistry, University of Texas, Austin, TX 78712

Toni Carbo Bearman (1994; *Section on Information, Computing, and Communication*), School of Library and Information Science, 505 LIS Bldg., University of Pittsburgh, 135 N. Bellefield St., Pittsburgh, PA 15260

Abram S. Benenson (1994; *Electorate on Medical Sciences*), Graduate School of Public Health, San Diego State University, San Diego, CA 92182

Daniel Berg (1994; *Electorate on Industrial Science*), CII 5015, Rensselaer Polytechnic Institute, 110 8th St., Troy, NY 12180-3590

Robert J. Birgeneau (1995; *Electorate on Physics*), School of Science, Rm. 6-123, Massachusetts Institute of Technology, Cambridge, MA 02139

Note: Terms end on the last day of the Annual Meeting held in the years given in parentheses.

C. Gunnar Blomqvist (1994; *Electorate on Medical Sciences*), Div. of Cardiology, H8.122, University of Texas Southwestern Medical Center, 5323 Harry Hines Blvd., Dallas, TX 75235-9034

Jill C. Bonner (1995; *Electorate on Physics*), Dept. of Physics, University of Rhode Island, Kingston, RI 02881

Edward N. Brandt, Jr. (1994; *Section on Medical Sciences*), Rm. 359, CHB, University of Oklahoma, HSC, P.O. Box 26901, Oklahoma City, OK 73190

Joost A. Businger (1994; *Section on Atmospheric and Hydrospheric Sciences*), P.O. Box 541, Anacortes, WA 98221

Edwin C. Cadman (1994; *Electorate on Medical Sciences*), Dept. of Internal Medicine, Yale University School of Medicine, 1072 LMP, 333 Cedar St., New Haven, CT 06510

Gloria V. Callard (1995; *Electorate on Biological Sciences*), Dept. of Biology, Boston University, 5 Cummington St., Boston, MA 02215

Byron A. Campbell (1994; *Electorate on Psychology*), Dept. of Psychology, Princeton University, Princeton, NJ 08544-1010

George R. Carruthers (1995; *Electorate on Astronomy*), Code 4109, Naval Research Lab., Washington, DC 20375-5000

Nancy Da Silva (1994; *Electorate on Engineering*), Biochemical Engineering Program, University of California, Irvine, CA 92717

Sharon Dunwoody (1994; *Section on General Interest in Science and Engineering*), School of Journalism and Mass Communication, University of Wisconsin, Madison, WI 53706

Patricia J. Eberlein (1995; *Electorate on Mathematics*), Dept. of Computer Science, 226 Bell Hall, State University of New York, Buffalo, NY 14260

William K. Estes (1994; *Section on Psychology*), Dept. of Psychology, Harvard University, 33 Kirkland St., Cambridge, MA 02138

David L. Featherman (1994; *Electorate and Section on Social, Economic, and Political Sciences*), Social Science Research Council, 605 Third Ave., New York, NY 10158

Susan Gottesman (1995; *Electorate on Biological Sciences*), Bldg. 37, Rm. 4B03, Lab. of Molecular Biology, National Cancer Institute, Bethesda, MD 20892

Susan L. Graham (1996; *Electorate on Information, Computing, and Communication*), Computer Science Div., EECS, University of California, Berkeley, CA 94720

John S. Greenspan (1994; *Section on Dentistry*), Dept. of Stomatology, Rm. S-612, University of California, San Francisco, CA 94143-0422

Karen A. Holbrook (1994; *Section on Biological Sciences*), Office of the Dean, SC-64, School of Medicine, University of Washington, Seattle, WA 98195

Rachelle D. Hollander (1996; *Electorate on Societal Impacts of Science and Engineering*), EVS, National Science Foundation, 1800 G St., N.W., Rm. 320, Washington, DC 20550

Edward L. Kaplan (1994; *Electorate on Medical Sciences*), Dept. of Pediatrics, Box 296, University of Minnesota Medical School, 420 Delaware St., S.E., Minneapolis, MN 55455

Alvin L. Kwiram (1994; *Section on Chemistry*), Office of Research, AH-20, University of Washington, Seattle, WA 98195

Nan M. Laird (1996; *Electorate on Statistics*), Dept. of Biostatistics, Harvard School of Public Health, 677 Huntington Ave., Boston, MA 02115

Judith A. Lengyel (1995; *Electorate on Biological Sciences*), Dept. of Biology, University of California, 405 Hilgard Ave., Los Angeles, CA 90024-1606

Alan E. Leviton (1995; *Pacific Division*), California Academy of Sciences, Golden Gate Park, San Francisco, CA 94118

Madeleine J. Long (1994; *Section on Education*), EHR, National Science Foundation, 1800 G St., N.W., Rm. 516, Washington, DC 20550

Jane Maienschein (1994; *Electorate on History and Philosophy of Science*), Dept. of Philosophy, Arizona State University, Tempe, AZ 85287-2004

Nancy H. Marcus (1995; *Electorate on Biological Sciences*), Dept. of Oceanography, Florida State University, Tallahassee, FL 32306

Ann G. Matthysse (1995; *Electorate on Biological Sciences*), Dept. of Biology, Coker Hall, CB 3280, University of North Carolina, Chapel Hill, NC 27599

Kurt Mislow (1995; *Electorate on Chemistry*), Dept. of Chemistry, Princeton University, Princeton, NJ 08544

Howard E. Morgan (1994; *Electorate on Medical Sciences*), Weis Center for Research, Geisinger Clinic, 100 N. Academy Ave., Danville, PA 17822-2601

Alice J. Moses (1996; *Electorate on Education*), ESIE, National Science Foundation, 1800 G St., N.W., Rm. 635A, Washington, DC 20550

Arno G. Motulsky (1994; *Electorate on Medical Sciences*), Div. of Medical Genetics, RG-25, University of Washington, Seattle, WA 98195

Donald J. Nash (1996; *Southwestern and Rocky Mountain Division*), Dept. of Biology, Colorado State University, Ft. Collins, CO 80523

John L. Neumeyer (1994; *Section on Pharmaceutical Sciences*), Research Biochemicals, Inc., 1 Strathmore Rd., Natick, MA 01760

Claire M. Oswald (1995; *National Association of Academies of Science*), Dept. of Biology, College of St. Mary, 1901 S. 72nd St., Omaha, NE 68124

Glenn Paulson (1994; *Section on Societal Impacts of Science and Engineering*), Dept. of Environmental Engineering, Illinois Institute of Technology, Chicago, IL 60616

Donald O. Pederson (1994; *Section on Engineering*), 1436 Via Loma, Walnut Creek, CA 94598

Calvin O. Qualset (1994; *Section on Agriculture*), Genetic Resources Conservation Program, University of California, Davis, CA 95616

David P. Rall (1994; *Electorate on Medical Sciences*), 5302 Reno Rd., N.W., Washington, DC 20015

George Rapp, Jr. (1995; *Electorate on Geology and Geography*), Archaeometry Lab., University of Minnesota, Duluth, MN 55812-2496

Richard J. Raridon (1994; *National Association of Academies of Science*), 111 Columbia Dr., Oak Ridge, TN 37830-7721

John D. Roberts (1995; *Electorate on Chemistry*), Crellin Lab. 164-30, California Institute of Technology, Pasadena, CA 91125

Nina M. Roscher (1995; *Electorate on Chemistry*), Dept. of Chemistry, American University, 4400 Massachusetts Ave., N.W., Washington, DC 20016-8014

Donald B. Rubin (1994; *Section on Statistics*), Dept. of Statistics, Harvard University, 1 Oxford St., Cambridge, MA 02138

Michael Ruse (1994; *Section on History and Philosophy of Science*), Dept. of Philosophy, University of Guelph, Guelph, Ont., Canada N1G 2W1

Alice T. Schafer (1994; *Section on Mathematics*), Dept. of Mathematics, Marymount University, Arlington, VA 22207-4299

John P. Schiffer (1994; *Section on Physics*), Physics Div., Bldg. 203, Argonne National Lab., 9700 S. Cass Ave., Argonne, IL 60439-4843

Kathleen L. Schroeder (1996; *Electorate on Dentistry*), Dept. of Oral Pathology, School of Dentistry, Health Sciences Center, West Virginia University, Morgantown, WV 26506

A. Richard Seebass (1994; *Electorate on Engineering*), Campus Box 422, University of Colorado, Boulder, CO 80309-0422

Sergio Silva-Ruiz (1994; *Caribbean Division*), Schering-Plough Products, Inc., P.O. Box 486, Manati, PR 00674-0486

Susan M. Simkin (1994; *Section on Astronomy*), Dept. of Physics and Astronomy, Michigan State University, E. Lansing, MI 48824-1116

Larry R. Squire (1994; *Electorate on Psychology*), V.A. Medical Center (116A), 3350 La Jolla Village Dr., San Diego, CA 92161

Diane K. Stoecker (1995; *Electorate on Biological Sciences*), Horn Point Environmental Lab., University of Maryland System, CEES, P.O. Box 775, Cambridge, MD 21613

Kevin E. Trenberth (1996; *Electorate on Atmospheric and Hydrospheric Sciences*), National Center for Atmospheric Research, 1850 Table Mesa Dr., P.O. Box 3000, ML428 CAS, Boulder, CO 80307-3000

Christy G. Turner, II (1994; *Section on Anthropology*), Dept. of Anthropology, Arizona State University, Tempe, AZ 85287-2402

Conrad J. Weiser (1994; *Electorate on Agriculture*), College of Agricultural Sciences, 137 Strand Agricultural Hall, Oregon State University, Corvallis, OR 97331-2212

Gunter Weller (1996; *Arctic Division*), Geophysical Institute, University of Alaska, Fairbanks, AK 99775-0800

Zena Werb (1995; *Electorate on Biological Sciences*), Lab. of Radiobiology and Environmental Health, LR-102, University of California, Box 0750, 3rd & Parnassus Aves., San Francisco, CA 94143-0750

Grant R. Wilkinson (1996; *Electorate on Pharmaceutical Sciences*), Dept. of Pharmacology, Vanderbilt University School of Medicine, Nashville, TN 37232-6600

Dael Wolfle (1996; *Electorate on General Interest in Science and Engineering*), Graduate School of Public Affairs, DC-13, University of Washington, Seattle, WA 98195

Jeannette Yen (1995; *Electorate on Biological Sciences*), Marine Sciences Research Center, State University of New York, Stony Brook, NY 11794-5000

Committees and Representatives

STANDING COMMITTEES

Council Affairs

Francisco J. Ayala (1996), Dept. of Ecology and Evolutionary Biology, University of California, Irvine, CA 92717, AAAS president-elect and **chair**

Eloise E. Clark (1995), Academic Affairs, 230 McFall Center, Bowling Green State University, Bowling Green, OH 43403, AAAS president

Daniel Berg (1994), Decision Sciences & Engineering Systems, CII 5015, Rensselaer Polytechnic Institute, 110 8th Street, Troy, NY 12180-3590

Jill C. Bonner (1995), Dept. of Physics, University of Rhode Island, Kingston, RI 02881

David L. Featherman (1994), Social Science Research Council, 605 Third Ave., New York, NY 10158

Judith A. Lengyel, (1995), Dept. of Biology, University of California, 405 Hilgard Ave., Los Angeles, CA 90024-1606

Jane Maienschein (1994), Philosophy Dept., University of Arizona, Tempe, AZ 85287-2004

Nancy H. Marcus (1995), Dept. of Oceanography, Florida State University, Tallahassee, FL 32306

George Rapp, Jr. (1995), Archaeometry Lab., University of Minnesota, Duluth, MN 55812-2496

Nina Matheny Roscher (1994), Dept. of Chemistry, American University, Washington, DC 20016-8014

Richard S. Nicholson, AAAS executive officer (ex officio)

Gretchen A. Seiler, AAAS staff officer

Executive Committee of the Board of Directors

F. Sherwood Rowland (1994), Dept. of Chemistry, University of California, Irvine, CA 92717, chairman of the board and **chair**

Eloise E. Clark (1995), Academic Affairs, 230 McFall Center, Bowling Green State University, Bowling Green, OH 43403, AAAS president

Francisco J. Ayala (1996), Dept. of Ecology and Evolutionary Biology, Room 716 Engineering Bldg., University of California, Irvine, CA 92717, AAAS president-elect

---

Note: Terms of committee members end on the last day of the Annual Meeting held in the year given in parentheses.

William T. Golden (1995), 40 Wall St., Rm. 4201, New York, NY 10005, AAAS Treasurer
Jean'ne Shreeve (1995), University Research Office, 111 Morrill Hall, University of Idaho, Moscow, ID 83843
Richard S. Nicholson, AAAS executive officer (ex officio)

Investment and Finance

Malcolm B. Smith (1995), General American Investors Co., Inc., 450 Lexington Ave., Ste. 3300, New York, NY 10017-3904, **chair**
Philip H. Abelson (1996), AAAS, 1333 H St., N.W., Washington, DC 20005
Mary Ellen Avery (1994), Harvard Medical School, 221 Longwood Ave., Boston, MA 02115
Richard Fischer (1994), Fischer, Francis, Trees & Watts, Inc., 717 Fifth Ave., New York, NY 10022
Robert G. Goelet (1994), Goelet Estate Co., 22 E. 67th St., New York, NY 10021
Helene L. Kaplan (1995), Skadden, Arps, Slate, Meagher & Flom, 919 Third Ave., New York, NY 10022-9931
William M. Kelly (1995), 40 Wall St., Rm. 4201, New York, NY 10005
Kevin W. Kennedy (1996), Goldman, Sachs & Co., 85 Broad St., New York, NY 10004.
John A. Levin (1996), John A. Levin & Co., Inc., One Rockefeller Plaza, 25th floor, New York, NY 10020.
Lloyd N. Morrisett (1996), John and Mary R. Markle Foundation, 75 Rockefeller Plaza, New York, NY 10019-6908
Eloise E. Clark (1994), Bowling Green State University, 230 McFall Center, Bowling Green, OH 43404, AAAS President (ex-officio)
William T. Golden, 40 Wall St., Rm. 4201, New York, NY 10005, AAAS Treasurer (ex officio)
Richard S. Nicholson, AAAS (ex officio)
Carl B. Amthor, AAAS staff officer

Nominations

David Korn (1994), Office of the Dean, School of Medicine, M-121, Stanford University, Stanford, CA 94305-5302
Julie Haynes Lutz (1994), Program in Astronomy, Washington State University, Pullman, WA 99164-3113
Kenneth R. Manning (1995), Rm. 20C-109, Massachusetts Institute of Technology, Cambridge, MA 02139
Joseph D. Novak (1995), Dept. of Education, 421 Kennedy Hall, Cornell University, Ithaca, NY 14853
Gordon H. Orians (1994), Department of Zoology, University of Washington, Seattle, WA 98195
Frank von Hippel (1995), Ctr. for Energy & Environmental Studies, H-102 Engineering Quadrangle., Princeton University, Princeton, NJ 08544

Marvalee H. Wake (1994), Department of Integrative Biology, University of California, Berkeley, CA 94720

Vivian Weil (1995), Ctr. for the Study of Ethics in the Professions, Illinois Institute of Technology, 3101 S. Dearborn St., Chicago, IL 60616-3793

Eloise E. Clark (1994), Academic Affairs, 230 McFall Center, Bowling Green State University, Bowling Green, OH 43403, AAAS president and board representative

Richard S. Nicholson, AAAS staff officer

COMMITTEES OF THE BOARD OF DIRECTORS

Annual Meeting Site Committee

F. Sherwood Rowland, Dept. of Chemistry, University of California, Irvine, CA 92717, chairman of the board (ex officio) and **chair**

Francisco J. Ayala, Dept. of Ecology and Evolutionary Biology, University of California, Irvine, CA 92717

Chang-Lin Tien, Office of the Chancellor, 200 California Hall, University of California, Berkeley, CA 94720

Richard S. Nicholson, AAAS executive officer (ex officio)

Robin Yeaton Woo, AAAS staff officer

Audit Committee

William T. Golden, 40 Wall St., Rm. 4201, New York, NY 10005, treasurer (ex officio) and **chair**

F. Sherwood Rowland, Dept. of Chemistry, University of California, Irvine, CA 92717, chairman of the board (ex officio)

Eloise E. Clark, Academic Affairs, 230 McFall Center, Bowling Green State University, Bowling Green, OH 43403, AAAS president

Alan Schriesheim, Argonne National Lab., 9700 South Cass Ave., Bldg. 201, Argonne, IL 60439-4832

Richard S. Nicholson, AAAS executive officer (ex officio)

Carl B. Amthor, AAAS staff officer

Compensation Committee

F. Sherwood Rowland, Dept. of Chemistry, University of California, Irvine, CA 92717, chairman of the board and **chair**

Eloise E. Clark, Academic Affairs, 230 McFall Center, Bowling Green State University, Bowling Green, OH 43403, AAAS president

Francisco J. Ayala, Dept. of Ecology and Evolutionary Biology, Rm 716 Engineering Bldg., University of California, Irvine, CA 92717, AAAS president-elect

William T. Golden, 40 Wall St., Rm. 4201, New York, NY 10005, AAAS treasurer

Robert A. Frosch, General Motors Research, General Motors Corporation, Warren, MI 48090-9055

Development Committee

Eloise E. Clark, Academic Affairs, 230 McFall Center, Bowling Green State University, Bowling Green, OH 43403, AAAS president (ex officio) and **chair**

Francisco J. Ayala, Dept. of Ecology and Evolutionary Biology, Rm. 716 Engineering Bldg., University of California, Irvine, CA 92717, AAAS president-elect

William T. Golden, 40 Wall St., Rm. 4201, New York, NY 10005, AAAS treasurer (ex officio)

Richard S. Nicholson, AAAS executive officer (ex officio)

Jeannette Wedel, AAAS staff officer

Health of the Scientific Enterprise Committee

Robert A. Frosch, General Motors Research, General Motors Corporation, Warren, MI 48090-9055, **chair**

William T. Golden, 40 Wall St., Rm. 4201 New York, NY 10005

Florence P. Haseltine, National Institutes of Health, Executive Plaza N., Rm. 604, Bethesda, MD 20892

Warren M. Washington, National Center for Atmospheric Research, P.O. Box 3000, Boulder, CO 80307-3000

Eloise E. Clark, Academic Affairs, 230 McFall Hall, Bowling Green State University, Bowling Green, OH 43403, AAAS president (ex officio)

Francisco J. Ayala, Dept. of Ecology and Evolutionary Biology, Room 716 Engineering Bldg., University of California, Irvine, CA 92717, AAAS president-elect (ex officio)

Long-Range Planning Committee

Eloise E. Clark, Academic Affairs, 230 McFall Center, Bowling Green State University, Bowling Green, OH 43403, AAAS president and **chair**

F. Sherwood Rowland, Dept. of Chemistry, University of California, Irvine, CA 92717, chairman of the board (ex officio)

Robert A. Frosch, General Motors Research, General Motors Corporation, Warren, MI 48090-9055

Warren M. Washington, National Center for Atmospheric Research, P.O. Box 3000, Boulder, CO 80307-3000

Richard S. Nicholson, AAAS executive officer (ex officio)

Prize Committee

Jean'ne Shreeve, University Research Office, 111 Morrill Hall, University of Idaho, Moscow, ID 83843, **chair**

Florence P. Haseltine, National Institutes of Health, Executive Plaza N., Rm. 604, Bethesda, MD 20892

Richard S. Nicholson, AAAS executive officer (ex officio)

Jeannette Wedel, AAAS staff officer

AWARD COMMITTEES

AAAS–Philip Hauge Abelson Prize

Selection Panel

James Baker (1994), Joint Oceanographics Institute, Inc., 1755 Massachusetts Ave., N.W., Washington, DC 20036
Floyd E. Bloom (1993), Research Institute of Scripps Clinic, 10666 N. Torrey Pines Rd., BCRI, La Jolla, CA 92037
Lewis M. Branscomb (1994), Program in Science, Technology and Public Policy, J.F.K. School of Government, Harvard University, Cambridge, MA 02138
William Brinkman (1995), AT&T Bell Laboratories, Physics Div., Room 1C-224, 600 Mountain Ave., Murray Hill, NJ 07974
Mary E. Clutter (1993), National Science Foundation, Biological, Behavioral and Social Sciences Directorate, Rm. 506 1800 G St., N.W., Washington, DC 20550
Robert F. Murray (1995), Howard University, Div. of Medical Genetics, College of Medicine, 520 W St., NW, PO Box 75, Washington, DC 20059
Rodney W. Nichols (1995), New York Academy of Sciences, 2 East 63rd St., New York, NY 10021
Stephen D. Nelson, AAAS staff officer

AAAS–Newcomb Cleveland Prize

Selection Committee (1991–1992)

Floyd E. Bloom, Preclinical Neuroscience, Scripps Clinic and Research Foundation, 10666 North Torrey Pines Road, La Jolla, CA 92037
Robert T. Schimke, Dept. of Biological Sciences, Stanford University, Stanford, CA 94305
Daniel E. Koshland, Jr., editor, *Science*
Richard S. Nicholson, AAAS executive officer
Philip H. Abelson, deputy editor, *Science,* and science advisor, AAAS
John I. Brauman, deputy editor, *Science*

AAAS Award for Behavioral Science Research

Selection Panel

G. David Johnson (1993), Dept. of Sociology, University of South Alabama, Mobile, AL 36688
Jerome Kagan (1993), Dept. of Psychology, Harvard University, William James Hall, 33 Kirkland St., Cambridge, MA 02138
Joseph Lopreato, (1993), Dept. of Sociology, The University of Texas at Austin, Austin, TX 78712-1088
Robert L. Munroe (1993), Dept. of Anthropology, Pitzer College, 201 Bernard Hall, Claremont, CA 91711
Earl Smith (1993), Dept. of Sociology, Pacific Lutheran University, Tacoma, WA 98447

Cookie White Stephan (1993), Dept. of Sociology/Anthropology, New Mexico State University, Las Cruces, NM 88003
Catherine Campos, AAAS staff officer

AAAS–Hilliard Roderick Prize

Selection Committee (1993)

Charles Zraket, John F. Kennedy School of Government, Harvard University, 79 John F. Kennedy St., Cambridge, MA 02138

M. Granger Morgan, Dept. of Engineering and Public Policy, Carnegie-Mellon University, 5000 Forbes Ave., Pittsburgh, PA 15213

Michael Krepon, Henry L. Stimson Center, Ste. 304, 1350 Connecticut Ave., N.W., Washington, DC 20036

Gloria Duffy, Global Outlook, 172 University Avenue, Palo Alto, CA 94301

W. Thomas Wander, Program on Science and International Security, AAAS, 1333 H Street, NW, Washington, DC 20005

AAAS Scientific Freedom and Responsibility Award

Selection Panel

Taft Broome, Jr. (1994), Civil Engineering Dept., Howard University, Washington, DC 20059

Mary-Claire King (1994), Dept. of Biomedical and Environmental Health Sciences, School of Public Health, University of California, Berkeley, CA 94720

Michael A. Lytle (1996), 5121 Hoag Lane, Fayetteville, NY 13066

Patricia Woolf (1996), Dept. of Sociology, Princeton University, Princeton, NJ 08544

Rosemary Chalk (1994), HA-172N, National Academy of Sciences, 2101 Constitution Ave., N.W., Washington, DC 20418

Deborah Runkle, AAAS staff officer

(One additional panel member to be selected.)

AAAS Award for Public Understanding of Science and Technology

Panel of Judges (1992)

Chris Raymond, Director of Publications, Association of Science-Technology Centers, 1025 Vermont Avenue NW, Washington, DC 20005

Alphonse Buccino, Office of Science and Technology Policy, Executive Office of the President, Washington, DC 20506

Joan Wrather, American Institute of Physics, 335 East 45th St., New York, NY 10017

Kim McDonald, Chronicle of Higher Education, 1255 23rd St. NW, Washington, DC

Marilyn Suiter, American Geological Institute, 4220 King St., Alexandria, VA 22302

AAAS–Westinghouse Science Journalism Awards

Managing Committee

Nan Broadbent, AAAS, 1333 H St., N.W., Washington, DC 20005

Ira Flatow, Samanna Productions, 45 Church St., Stamford, CT 06906
Denise Graveline, AAAS, 1333 H St., N.W., Washington, DC 20005
Ronald Hart, Westinghouse Electric Corporation, Gateway Center, Pittsburgh, PA 15222
Kim McDonald, *Chronicle of Higher Education*, 1255 23rd St., N.W., Suite 600, Washington, DC 20037
Paul Recer, Associated Press, 2021 K St., N.W., Washington, DC 20006

Print Screening Committee

Rick Borchelt, Committee on Science, Space, and Technology, U. S. House of Representatives, 2320 RHOB, Washington, DC 20515
Sharon Dunwoody, University of Wisconsin-Madison, School of Journalism and Mass Communications, Madison, WI 53706
Ann Gibbons, *Science*, 1333 H St., N.W., Washington, DC 20005
Peggy Girshman, National Public Radio, 2025 M St. N.W., Washington, DC 20036
Steve Maran, NASA, GSSC, Code 680, Greenbelt, MD 20771
Rob Stein, United Press International, 1400 Eye St., N.W., Washington, DC 20005

Print Judging Committee

Graeme Browning, *National Journal*, 1730 M St., N.W., #1100, Washington, DC 20036
Celia Hooper, *Journal of NIH Research*, 2101 L St., N.W., Suite. 207, Washington, DC 20006
Robert Lee Hotz, *Atlanta Journal-Constitution*, P.O. Box 4689, Atlanta, GA 30302
Victor K. McElheny, Knight Science Journalism Fellowships, M.I.T., 113 Huron Ave., Cambridge, MA 02138
Dennis Meredith, Office of Research Communications, Duke University, 615 Chapel Drive, Durham, NC, 27706
Beverly Orndorff, *Richmond Times-Dispatch*, P.O. Box 85333, Richmond, VA 23293
Rita Rubin, *U. S. News and World Report*, 2400 N St., N.W., Washington, DC 20037
David Voss, *Science*, 1333 H St., N.W., Washington, DC 20005

Radio Judging Committee

Bob Hirshon, AAAS Radio, 1333 H St., N.W., Washington, DC 20005
Craig Katz, Mutual Radio, 1755 S. Jefferson Davis Hwy., Arlington, VA 22202
Richard Kerr, *Science*, 1333 H St., N.W., Washington, DC 20005
Doug Levy, United Press International, 1400 Eye St., N.W., Washington, DC 20005

Television Screening

Carol Chaney, Husson College, Bangor Maine 04401
Virgil Frizzell, U. S. Geological Survey, 12201 Sunrise Valley Dr., Reston, VA 22092
Gilbert Herrera, Sandia National Laboratories, P. O. Box 5800, Albuquerque, NM 87114
Van Sickler, U. S. Food and Drug Administration, 5600 Fishers Lane, Rockville, MD 20857
Peter Zimmerman, Center for Strategic & International Studies, 1800 K St. N.W., Suite 400, Washington, DC 20006

Television Judging Committee

Maura Lerner, *Minneapolis Star & Tribune*, 425 Portland Ave., Minneapolis, MN 55488
Gordon Rothman, "CBS This Morning," 524 W. 57th St., New York, NY 10019
Don Torrance, Newhouse School of Publications, Syracuse University, Newhouse II, 215 University Place, Syracuse, NY 13244-2100

AAAS Mentor Award

Last year's award was selected by the AAAS Committee on Opportunities in Science.

AAAS International Scientific Cooperation Award

To be awarded for the first time in 1993.

BOARD–APPOINTED COMMITTEES*

AAAS–ABA National Conference of Lawyers and Scientists

Richard A. Meserve (1994), Covington and Burling, 1201 Pennsylvania Ave., N.W., P.O. Box 7566, Washington, DC 20044, **co-chair**
Lee Loevinger (1993), Hogan & Hartson, Columbia Sq., 555 13th St., N.W., Washington, DC 20004, **co-chair**
R. Stephen Berry (1995), Dept. of Chemistry, University of Chicago, 5735 S. Ellis Ave., Chicago, IL 60637
Bert Black (1996), Weinberg and Green, 100 S. Charles St., Baltimore, MD 21201
Linda M. Distlerath (1996), Merck & Co., Inc., One Merck Dr., P.O. Box 100, Mail Code WS1A-28, Whitehouse Station, NJ 08889
Andrea Bear Field (1993), Hunton & Williams, 2000 Pennsylvania Ave., N.W., Washington, DC 20006
Alan I. Leshner (1994), NIMH/National Institutes of Health, 5600 Fishers La., Rockville, MD 20857
Harold Lurie (1995), California Council on Science and Technology, Arnold & Mabel Beckman Ctr., National Academies of Science and Engineering, 100 Academy Dr., Irvine, CA 92715
Barbara Mishkin (1995), Hogan & Hartson, Columbia Sq., 555 13th St., N.W., Washington, DC 20004
Susan H. Nycum (1994), Baker & McKenzie, 660 Hansen Way, Palo Alto, CA 94304
Gilbert S. Omenn (1944), School of Public Health and Community Medicine, University of Washington, Seattle, WA 98195
John F. Shoch (1994), Asset Management Co., 2275 E. Bayshore, Palo Alto, CA 94303
Oliver R. Smoot (1994), Computer and Business Equipment Manufacturers Association, 1250 Eye St., N.W., Ste. 200, Washington, DC 20005
Nicholas C. Yost (1994), Dickstein, Shapiro & Morin, 2101 L St., N.W., Ste. 800, Washington, DC 20037
Ruth C. Burg, liaison, ABA Standing Committee on National Conference Groups, 7th Fl., Skyline 6, 5109 Leesburg Pike, Falls Church, VA 22041

---

* The executive officer serves as an ex officio member of these committees.

Robin Roy, staff liaison, American Bar Assn., 750 N. Lake Shore Dr., Chicago, IL 60611
Florence P. Haseltine, National Institutes of Health, Executive Plaza N., Rm. 604, Bethesda, MD 20892, board liaison
Mark S. Frankel, AAAS staff officer
Deborah Runkle, AAAS associate staff officer

Annual Meeting Scientific Program Committee

Eloise E. Clark, Academic Affairs, Bowling Green State University, 230 McFall Ctr., Bowling Green, OH 43403, **chair**
Francisco J. Ayala, Dept. of Ecology and Evolutionary Biology, University of California, Irvine, CA 92717
James Bower, Div. of Biology 216-76, California Institute of Technology, Pasadena, CA 91125
Ernestine Friedl, Dept. of Cultural Anthropology, Duke University, Durham, NC 27706
Ron Graham, AT&T Bell Labs, 600 Mountain Ave., Murray Hill, NJ 07974
Judith Tegger Kildow, Dept. of Ocean Engineering, Massachusetts Institute of Technology, 5-214, Cambridge, MA 02139
Donald A.B. Lindberg, National Library of Medicine, MLN Bldg. 38, Rm 2E, 8600 Rockville Pike, Bethesda, MD 20894
Orie L. Loucks, Dept. of Zoology, Miami University, Oxford, OH 45056
Cora Bagley Marrett, National Science Foundation, Directorate for Social, Behavioral and Economic Studies, Washington, DC 20550
Robert Morgan, Dept. of Engineering & Policy, Washington University, Box 1106, One Brookings Dr., St. Louis, MO 63130
Gilbert S. Omenn, School of Public Health SC-30, University of Washington, Seattle, WA 98195
Stuart A. Rice, James Franck Institute, University of Chicago, 5640 Ellis Ave., Chicago, IL 60637
Gordon Wolman, Dept. of Geography and Environmental Engineering, Johns Hopkins University, Baltimore, MD 21218
Harry Woolf, Institute for Advanced Study, Olden Ln., Princeton, NJ 08540
Robin Yeaton Woo, AAAS Staff Officer

Opportunities in Science

Manuel Gomez-Rodriquez (1995), Puerto Rico Resource Center for Science and Engineering, College of Natural Sciences, University of Puerto Rico, Secundo Bueso, Ofc. #304, Rio Piedras, PR 00931, **chair**
Margarita Colmenares (1994), Chevron International Oil Company, Inc., 555 Market Street, P.O. Box 7148, San Francisco, CA 94120-7148
Kent Cullers (1996), SETI Program Office, NASA, Ames Research Center, MS N244-11, Moffett Field, CA 94035
Freeman A. Hrabowski, III (1996), Interim President, University of Maryland, Baltimore County, Baltimore, MD 21228
Jane Kahle (1994), Condit Professor of Science Education, 418 McGuffey Hall, Miami University, Oxford, OH 45056

Willie Pearson, Jr. (1995), P.O. Box 808, Department of Sociology, Wake Forest University, Winston-Salem, NC 21099
Karl S. Pister (1995), Interim Chancellor, 296 McHenry Library, University of California, Santa Cruz, CA 95064
Gloria Scott (1994), President, Bennett College, 900 E. Washington St., Greensboro, NC 27401-3239
Elizabeth K. Stage (1996), National Academy of Science, 2101 Constitution Avenue, N.W., HA-486, Washington, D.C. 20418
Sylvia Walker (1994), Howard University Research and Training Center for Access to Rehabilitation and Economic Opportunity, 2900 Van Ness St., N.W., Washington, DC 20008
Warren M. Washington, National Center for Atmospheric Research, P.O. Box 3000, Boulder, CO 80307-3000 board representative
William A. Lester, Jr., Dept. of Chemistry, University of California, Berkeley, CA 94720, board representative
Yolanda D. George, AAAS staff officer

Public Understanding of Science and Technology

Marcel C. LaFollette (1996), Center for International Science and Technology Policy, 2130 H St., N.W., Ste. 714, The George Washington University, Washington, DC 20052, **chair**
John Angier (1995), The Chedd-Angier Production Company, 70 Coolidge Hill Rd., Watertown, MA 02172
Edward Atkins (1996), Children's Television Workshop, 1 Lincoln Plaza, New York, NY 10023
Valerie Crane (1994), Research Communications, 824 Boyleston St., Chestnut, MA 02167
David L. Crippens (1994), Educational Enterprises, KCET-TV, 4401 Sunset Blvd., Los Angeles, CA 90028
Sharon Dunwoody (1995), School of Journalism and Mass Communication, University of Wisconsin–Madison, Madison, WI 53706
Paul Hoffman (1994), *Discover,* 114 5th Ave., New York, NY 10011
Bruce V. Lewenstien (1995), Depts. of Communications and Science and Technology Studies, 321 Kennedy Hall, Cornell University, Ithaca, NY 14853
Alan McGowan, (1996) Scientists' Institute for Public Information, 355 Lexington Avenue, New York, NY 10017
Catherine Morrison (1995), Families and Work Institute, 330 7th Ave., New York, NY 10001
Talbert Spence (1996), American Museum of Natural History, Central Park West at 79th St., New York, NY 10024
Hill Williams (1995), 17263 Greenwood Place North, Seattle, WA 98133
Jean'ne M. Shreeve, University Research Office, 111 Morrill Hall, University of Idaho, Moscow, ID 83843, board representative
Robert A. Frosch, General Motors Research Lab., General Motors Corporation, Warren, WI 48090-9055, board representative

Nancy S. Wexler, Depts. of Neurology & Psychiatry, College of Physicians and Surgeons of Columbia University, 722 West 168th St., New York, NY 10033, board representative
Judy Kass, AAAS staff officer

Science and International Security

Sidney N. Graybeal (1994), Center for National Security Negotiations, SAIC, 1710 Goodridge Dr., McLean, VA 22102, **chair**
Paul Bracken (1995), Yale School of Organization and Management, 135 Prospect St., New Haven, CT 06511
Ashton B. Carter (1994), John F. Kennedy School of Government, Harvard University, 79 John F. Kennedy St., Cambridge, MA 02138
Stephen Cohen (1995), Dept. of Political Science, University of Illinois, 361 Lincoln Hall, 702 S. Wright St., Urbana, IL 61801
Alton Frye (1994), Council on Foreign Relations, 2400 N St., N.W., Washington, DC 20037
Peter H. Gleick, (1996), Global Environment Program, Pacific Institute for Studies in Development, Environment, and Security, 1204 Preservation Park Way, Oakland, CA 94612
Thomas Homer-Dixon, (1996), Peace and Conflict Studies, University of Toronto, Toronto, Canada, M5S 1A1
Kent H. Hughes (1994), Council on Competitiveness, 900 17th Street, N.W., Ste. 1050, Washington, DC 20006
Catherine M. Kelleher (1994), Foreign Policy Studies Program, The Brookings Institution, 1775 Massachusetts Ave., N.W., Washington, DC 20036-2188
Matthew Meselson (1996), Dept. of Biochemistry and Molecular Biology, Harvard University, 7 Divinity Ave., Cambridge, MA 02138
M. Granger Morgan (1994), Dept. of Engineering and Public Policy, Carnegie Mellon University, 5000 Forbes Ave., Pittsburgh, PA 15213
Janne E. Nolan (1995), Foreign Policy Studies Program, The Brookings Institution, 1775 Massachusetts Ave., N.W., Washington, DC 20036-2188
Amy Sands (1995), Lawrence Livermore National Lab., International Assessments Section, Z Division, L-389, 7000 East Ave., Livermore, CA 94551
Charles A. Zraket (1996), John F. Kennedy School of Government, Harvard University, 79 John F. Kennedy St., Cambridge, MA 02138
William T. Golden, 40 Wall St., Rm. 4201, New York, NY 10005, board liaison
W. Thomas Wander, AAAS staff officer

Science, Engineering, and Public Policy

Roberta Balstad Miller (1994), Consortium for International Earth Science Information Network, 2250 Pierce Road, University Center, MI 48710, **chair**
George Campbell, Jr. (1996), National Action Council for Minorities in Engineering, Inc., 3 West 35th St., New York, NY 10001
Nancy Carson (1994), Office of Technology Assessment, U.S. Congress, 600 Pennsylvania Ave., S.E., Washington, DC 20510-8025

William C. Clark (1994), Center for Science and International Affairs, Harvard University, 79 Kennedy St., Cambridge, MA 02138
Irwin Feller (1996), Institute for Policy Research and Evaluation, N253 Burrowes Bldg., Pennsylvania State University, University Park, PA 16802
Craig I. Fields (1995), Microelectronics and Computer Technology Corporation (MCC), 3500 W. Balcones Center Dr., Austin, TX 78759-6509
S. Allen Heininger (1995), 7811 Carondelet, Ste. 308, Clayton, MO 63105
Christopher Hill (1995), Critical Technologies Institute, RAND Corp., 2100 M St., N.W., Washington, DC 20037
C. Judson King, Provost, Professional Schools and Colleges, University of California, Berkeley, CA 94720
Thomas E. Malone (1995), Association of American Medical Colleges, 2450 N St., N.W., Washington, DC 20037-1126
J. David Roessner (1995), School of Public Policy, Georgia Institute of Technology, Atlanta, GA 30332
Frances E. Sharples (1996), Environmental Sciences Div., Oak Ridge National Lab., P.O. Box 2008, Oak Ridge, TN 37831-6036
Jon M. Veigel (1995), Oak Ridge Associated Universities, P.O. Box 117, Oak Ridge, TN 37831-0117
Francisco J. Ayala, Dept. of Ecology and Evolutionary Biology, University of California, Irvine, CA 92717, board liaison
Alan Schriesheim, Argonne National Lab., 9700 S. Cass Ave., Argonne, IL 60439-4832, board representative
Chang-Lin Tien, Office of Chancellor, University of California, 200 California Hall, Berkeley, CA 94720
Richard S. Nicholson, AAAS (ex officio)
Stephen D. Nelson, AAAS staff officer

Scientific Freedom and Responsibility

C.K. Gunsalus (1994), University of Illinois at UrbanaChampaign, 4th Fl. Swanlund Bldg., 601 E. John St., Champaign, IL 61820, **chair**
Alan Beyerchen (1994), Dept. of History, Ohio State University, 230 W. 17th Ave., Columbus, OH 43210
Katherine L. Bick (1996), 7300 Greentree Rd., Bethesda, MD 20817
John H. Bodley (1994), Dept. of Anthropology, Washington State Univ., Pullman, WA 99164
Nina Byers (1995), Dept. of Physics, University of California, Los Angeles, CA 90024
Richard P. Claude (1995), 3225 Grace St., N.W., Washington, DC 20007
Clarence J. Dias (1995), International Center for Law in Development, 777 United Nations Plz., Ste. 7E, New York, NY 10017
Rosa E. Garcia-Peltoniemi (1996), Center for Victims of Torture, 717 E. River Rd., Minneapolis, MN 55455
Mary W. Gray (1996), Dept. of Mathematics & Statistics, American University, 4400 Massachusetts Ave., N.W., Washington, DC 20016
Jane Bortnick Griffith (1994), Science Policy Research Div., Congressional Research

Service, Library of Congress, Washington, DC 20540
John Ladd (1995), Dept. of Philosophy, Box 1918, Brown University, Providence, RI 02912
William W. Middleton (1995), 6 Cornwall Cir., St. Davids, PA 19087
Drummond Rennie (1996), Institute for Health and Policy Studies, University of California, 1388 Sutter St., 11th Fl., San Francisco, CA 94109
Caroline Whitbeck (1995), Mechanical Engineering, Massachusetts Institute of Technology, MIT 1-104A, Cambridge, MA 02139
Mark S. Frankel, AAAS staff officer
Audrey R. Chapman, AAAS staff officer, human rights

REPRESENTATIVES TO OTHER ORGANIZATIONS

American Association for Accreditation of Laboratory Animal Care
Richard G. Traystman, Anesthesiology/Critical Care Medicine, The Johns Hopkins University, Baltimore, MD 21287

AAAS Commissioners to the Commission on Professionals in Science and Technology
Alan Fechter, Office of Scientific and Engineering Personnel, National Academy of Sciences, 2101 Constitution Ave., N.W., Washington, DC 20418
Nina Matheny Roscher, Dept. of Chemistry, American University, Washington, DC 20016
Shirley M. Malcom, AAAS staff liaison

Gordon Research Conferences Board of Trustees
Richard S. Nicholson, AAAS, 1333 H St., N.W., Washington, DC 20005

Interciencia Association
Helen Thomas, AAAS, 1333 H St., N.W., Washington, DC 20005, **acting executive director**
Leonard M. Rieser, Dickey Endowment, 207 Baker Library, Dartmouth College, Hanover, NH 03755
William Sawyer, China Medical Board, 750 Third Ave., 23rd Floor, New York, NY 10017
Barbara A. Timmerman, Dept. of Pharmaceutical Sciences, Health Sciences Center, University of Arizona, Tucson, AZ 85721

National Selection Committee
National Inventors Hall of Fame
Robert H. Rines, Franklin Pierce College Law Center, Concord, NH 03301

Science Editorial Board

Daniel E. Koshland, Jr., AAAS, 1333 H St., N.W., Washington, DC 20005, **chair**
Charles J. Arntzen, Institute of Biosciences and Technology, 2121 Holcombe Blvd., Houston, TX 77030
Elizabeth E. Bailey, Dept. of Public Policy and Management, 3107 SH-DH, University of Pennsylvania, 3620 Locust Walk, Philadelphia, PA 19104-6372
David Baltimore, The Rockefeller University, 1230 York Ave., New York, NY 10021
William F. Brinkman, Research, Physics Div. (Rm. 1C224), AT&T Bell Labs, 600 Mountain Ave., Murray Hill, NJ 07974
E. Margaret Burbidge, Center for Astrophysics and Space Sciences, Mail Code C-011, University of California–San Diego, La Jolla, CA 92093
Pierre-Gilles de Gennes, College de France, École Supérieure de Physique et de Chimie Industrielles de la Ville de Paris, 10, Rue Vauquelin, 75231 Paris Cedex 05, France
Joseph L. Goldstein, Dept. of Molecular Genetics, University of Texas Southwestern Medical Center, 5323 Harry Hines Blvd., Dallas, TX 75235
Mary L. Good, Allied-Signal, Inc., CTC Bldg., Columbia Rd. and Park Ave., Morristown, NJ 07960
Harry B. Gray, Beckman Institute and Div. of Chemistry and Chemical Engineering, California Institute of Technology, Pasadena, CA 91125
John J. Hopfield, Div. of Chemistry and Biology, California Institute of Technology, Pasadena, CA 91125
F. Clark Howell, Dept. of Anthropology, University of California, Berkeley, CA 94720
Paul A. Marks, Memorial Sloan-Kettering Cancer Center, 1275 York Ave., New York, NY 10021
Yasutomi Nishizuka, Dept. of Biochemistry, Kobe University School of Medicine, Kobe 650, Japan
Helen M. Ranney, Alliance Pharmaceutical Corporation, 3040 Science Park Rd., San Diego, CA 92122
Robert M. Solow, Dept. of Economics, Massachusetts Institute of Technology, Cambridge, MA 02139
Edward C. Stone, Jet Propulsion Lab., California Institute of Technology, 4800 Oak Grove Dr., Pasadena, CA 91109-8099
James D. Watson, Cold Spring Harbor Lab., P.O. Box 100, Cold Spring Harbor, NY 11724
Francisco J. Ayala, Dept. of Ecology and Evolutionary Biology, University of California, Irvine, Irvine, CA 92717, AAAS president-elect and board representative

Sections: Officers and Committee Members

The affairs of each of the Association's sections are managed by a section committee consisting of the chair, chair-elect, retiring chair, and secretary of the section; four members-at-large; Council delegate(s) of the electorate; and representatives of the affiliated organizations enrolled in the section. The section committees of each of the 23 sections are listed below. Each section committee meets during the Annual Meeting; between meetings, a section steering group (section officers and members-at-large) is responsible for section affairs.

AGRICULTURE

Officers

Chair: Kenneth J. Frey, Dept. of Agronomy, Iowa State University, Ames, IA 50011
Chair-Elect: Robert F Barnes, American Society of Agronomy, 677 S. Segoe Rd., Madison, WI 53711
Retiring Chair: Calvin O. Qualset, Genetic Resources Conservation Program, University of California, Davis, CA 95616
Secretary: E. C. A. Runge (1996), Soil & Crop Science Dept., Texas A&M University, College Station, TX 77843-2474

Members-at-Large

David V. Glover (1994), Dept. of Agronomy, Purdue University, W. Lafayette, IN 47907
Larry E. Schrader (1995), College of Agriculture and Home Economics, Washington State University, Pullman, WA 99164-6242
Eldon E. Ortman (1996), Agricultural Research, 1140 Ag. Administration Bldg., Purdue University, W. Lafayette, IN 47906-1140
Sue A. Tolin (1997), Dept. of Plant Pathology, Physiology, and Weed Science, Virginia Polytechnic Institute and State University, Blacksburg, VA 24061-0330

Council Delegate

Conrad J. Weiser (1994), College of Agricultural Sciences, 137 Strand Agricultural Hall, Oregon State University, Corvallis, OR 97331-2212

Representatives of Affiliates

Robert F Barnes (1995; *American Society of Agronomy*), American Society of Agronomy, 677 S. Segoe Rd., Madison, WI 53711

Note: Terms end on the last day of the Annual Meeting held in the years given in parentheses.

Vernon B. Cardwell (1996; *Crop Science Society of America*), Dept. of Agronomy, University of Minnesota, 1991 Buford Cir., St. Paul, MN 55108

Jimmy H. Clark (1995; *American Dairy Science Association*), 315 ASL, University of Illinois, 1207 W. Gregory Dr., Urbana, IL 61801

Charlie G. Coble (1996; *American Society of Agricultural Engineers*), Dept. of Agricultural Engineering, Texas A&M University, College Station, TX 77843

Charles H. Harden (1994; *Society of American Foresters*), Society of American Foresters, 5400 Grosvenor La., Bethesda, MD 20814

James Henderson (1995; *Alabama Academy of Science*), Carver Research Lab., Tuskegee University, Tuskegee, AL 36088

Scott Hutchins (1996; *Entomological Society of America*), Dowelanco, P.O. Box 681428, Indianapolis, IN 46268-7428

John Patrick Jordan (1995; *American Institute of Biological Sciences*), USDA, CSRS, Rm. 305A, Administration Bldg., Washington, DC 20250

Dennis R. Keeney (1994; *Soil and Water Conservation Society*), Leopold Center for Sustainable Agriculture, 126 National Soil Tilth Lab., Iowa State University, Ames, IA 50011-1010

Peter F. Korsching (1995; *Rural Sociological Society*), Dept. of Sociology, Iowa State University, Ames, IA 50011

Brijeshwar D. Mathur (1994; *Volunteers in Technical Assistance*), VITA, Inc., 1815 N. Lynn St., Ste. 200, Arlington, VA 22209-8438

Nell I. Mondy (1996; *Sigma Delta Epsilon, Graduate Women in Science*), 126 Honness La., Ithaca, NY 14850

Steven C. Nelson (1995; *American Association of Cereal Chemists* and *American Phytopathological Society*), American Assn. of Cereal Chemists, 3340 Pilot Knob Rd., St. Paul, MN 55121

Michael D. Ruff (1995; *Poultry Science Association*), LPSI, Bldg. 1040, BARC-East, Beltsville, MD 20705

Daniel R. Tompkins (1995; *American Society for Horticultural Science*), USDA, CSRS, Aerospace Bldg., Washington, DC 20250

Ross M. Welch (1995; *Soil Science Society of America*), U.S. Plant, Soil, and Nutrition Lab., Tower Rd., Cornell University, Ithaca, NY 14853

Robert G. Zimbelman (1996; *American Society of Animal Science*), FASEB, 9650 Rockville Pike, Bethesda, MD 20814

ANTHROPOLOGY

Officers

Chair: Frank Hole, Dept. of Anthropology, Yale University, Box 2114, Yale Station, New Haven, CT 06520

Chair-Elect: Jeremy A. Sabloff, Dept. of Anthropology, University of Pittsburgh, Pittsburgh, PA 15260

Retiring Chair: Christy G. Turner, II, Dept. of Anthropology, Arizona State University, Tempe, AZ 85287-2402

Secretary: Anna C. Roosevelt (1997), Dept. of Anthropology, Field Museum of Natural History, Lake Shore Dr. & Roosevelt Rd., Chicago, IL 60605-2496

Members-at-Large

Allen Johnson (1994), Dept. of Anthropology, University of California, Los Angeles, CA 90024

William A. Longacre (1995), Dept. of Anthropology, University of Arizona, Tucson, AZ 85721

Margaret J. Schoeninger (1996), Dept. of Anthropology, University of Wisconsin, Madison, WI 53706

Wendy Ashmore (1997), Dept. of Anthropology, 325 University Museum, University of Pennsylvania, Philadelphia, PA 19104-6398

Council Delegate

George J. Armelagos (1995), Dept. of Anthropology, Emory University, Atlanta, GA 30322

Representatives of Affiliates

Patricia Gindhart (1995; *Anthropological Society of Washington*), Dept. of Anthropology, American University, 4400 Massachusetts Ave., N.W., Washington, DC 20016-8003

Rebecca Huss-Ashmore (1996; *Human Biology Council*), Dept. of Anthropology, University Museum, University of Pennsylvania, Philadelphia, PA 19104-6398

Stephen C. Jett (1996; *Association of American Geographers*), Dept. of Geography, University of California, Davis, CA 95616

George T. Jones (1994; *Society for American Archaeology*), Dept. of Anthropology, Hamilton College, Clinton, NY 13323

Solomon H. Katz (1995; *American Association of Physical Anthropologists*), Dept. of Anthropology, University Museum, Rm., 341, University of Pennsylvania, Philadelphia, PA 19104

Stephen P. Koob (1994; *Archaeological Institute of America*), Freer Gallery of Art, Smithsonian Institution, Washington, DC 20560

Emilio F. Moran (1994; *Society for Applied Anthropology*), Dept. of Anthropology, Rawles Hall 108, Indiana University, Bloomington, IN 47405

Hart Nelsen (1994; *Society for the Scientific Study of Religion*), Dept. of Sociology, Pennsylvania State University, University Park, PA 16802

Michael Silverstein (1996; *Linguistic Society of America*), Dept. of Anthropology, University of Chicago, 1126 E. 59th St., Chicago, IL 60637

Clyde C. Snow (1995; *American Academy of Forensic Sciences*), 2230 Blue Creek Pkwy., Norman, OK 73071

Norman E. Whitten, Jr. (1995; *American Anthropological Association* and *American Ethnological Society*), Dept. of Anthropology, University of Illinois, Urbana, IL 61801

ASTRONOMY

Officers

Chair: Hugh M. Van Horn, Dept. of Physics and Astronomy, University of Rochester, Rochester, NY 14627-0011
Chair-Elect: Julie Haynes Lutz, Program in Astronomy, Washington State University, Pullman, WA 99164-3113
Retiring Chair: Susan M. Simkin, Dept. of Physics and Astronomy, Michigan State University, E. Lansing, MI 48824-1116
Secretary: Nancy Houk (1997), Dept. of Astronomy, 1041 Dennison Bldg., University of Michigan, Ann Arbor, MI 48109-1090

Members-at-Large

Robert D. Gehrz (1994), Dept. of Astronomy, University of Minnesota, 116 Church St., S.E., Minneapolis, MN 55455
Robert M. Hjellming (1995), National Radio Astronomy Observatory, P.O. Box 0, Socorro, NM 87801-0379
Martha L. Hazen (1996), Harvard College Observatory, 60 Garden St., Cambridge, MA 02138
Harry L. Shipman (1997), Dept. of Physics and Astronomy, University of Delaware, Newark, DE 19716

Council Delegate

George R. Carruthers (1995), Code 4109, Naval Research Lab., Washington, DC 20375-5000

Representatives of Affiliates

Marc S. Allen (1995; *American Astronautical Society*), National Research Council, 2101 Constitution Ave., N.W., Harris 584, Washington, DC 20418
Peter B. Boyce (1995; *American Institute of Physics*), American Astronomical Society, 2000 Florida Ave., N.W., Ste. 300, Washington, DC 20009
Louis Friedman (1994; *The Planetary Society*), The Planetary Society, 65 N. Catalina Ave., Pasadena, CA 91106
Christine Jones (1995; *American Astronomical Society*), Center for Astrophysics, 60 Garden St., Cambridge, MA 02138
P. Kenneth Seidelmann (1996; *Institute of Navigation*), U.S. Naval Observatory, 34th & Massachusetts Ave., N.W., Washington, DC 20390

ATMOSPHERIC AND HYDROSPHERIC SCIENCES

Officers

Chair: John Firor, Advanced Study Program, National Center for Atmospheric Research, P.O. Box 3000, Boulder, CO 80307-3000

Chair-Elect: Ferdinand Baer, Dept. of Meteorology, University of Maryland, College Park, MD 20742
Retiring Chair: Joost A. Businger, P.O. Box 541, Anacortes, WA 98221
Secretary: William H. Beasley (1995), School of Meteorology, 1310 Energy Center, University of Oklahoma, Norman, OK 73019

Members-at-Large

Stephen H. Schneider (1994), Dept. of Biological Sciences, Gilbert Bldg., Stanford University, Stanford, CA 94305-5020
Michael C. MacCracken (1995), Lawrence Livermore National Lab., P.O. Box 808 (L-262), Livermore, CA 94551-9900
John A. Dutton (1996), College of Earth and Mineral Sciences, 116 Deike Bldg., Pennsylvania State University, University Park, PA 16802
Jennifer A. Logan (1997), 108 Pierce Hall, Harvard University, 29 Oxford St., Cambridge, MA 02138

Council Delegate

Kevin E. Trenberth (1996), National Center for Atmospheric Research, 1850 Table Mesa Dr., P.O. Box 3000, ML428 CAS, Boulder, CO 80307-3000

Representatives of Affiliates

Richard Barber (1996; *American Society of Limnology and Oceanography*), Duke University Marine Lab., Beaufort, NC 28516
Martin J. Finerty, Jr. (1996; *Marine Technology Society*), Marine Technology Society, 1828 L St., N.W., Ste. 906, Washington, DC 20036-5104
Louis Friedman (1994; *The Planetary Society*), The Planetary Society, 65 N. Catalina Ave., Pasadena, CA 91106
Richard E. Hallgren (1995; *American Meteorological Society*), American Meteorological Society, 45 Beacon St., Boston, MA 02108
David S. Johnson (1995; *American Astronautical Society*), 1133 Lake Heron Dr., Apt. 3A, Annapolis, MD 21403
John R. Mather (1995; *American Geographical Society*), Dept. of Geography, University of Delaware, Newark, DE 19716
Linda O. Mearns (1996; *Association of American Geographers*), National Center for Atmospheric Research, P.O. Box 3000, Boulder, CO 80307-3000
Christopher N. K. Mooers (1995; *American Geophysical Union*), Div. of Applied Marine Physics, University of Miami, RSMAS, 4600 Rickenbacker Causeway, Miami, FL 33149-1098
John G. Weihaupt (1994; *Geological Society of America*), Dept. of Geology, University of Colorado, 1200 Larimer St., Box 172, Denver, CO 80204
Warren W. Wood (1995; *American Geological Institute*), U.S. Geological Survey, National Center, M.S. 431, Reston, VA 22092

BIOLOGICAL SCIENCES

Officers

Chair: Francisco J. Ayala, Dept. of Ecology and Evolutionary Biology, University of California, Irvine, CA 92717
Chair-Elect: Rita R. Colwell, Maryland Biotechnology Institute, 1123 Microbiology Bldg., University of Maryland, College Park, MD 20742
Retiring Chair: Karen A. Holbrook, Office of the Dean, SC-64, School of Medicine, University of Washington, Seattle, WA 98195
Secretary: Marjorie L. Reaka-Kudla (1997), Dept. of Zoology, University of Maryland, College Park, MD 20742

Members-at-Large

Juliana C. Mulroy (1994), Dept. of Biology, Denison University, Granville, OH 43023
Sarah C. R. Elgin (1995), Dept. of Biology, Box 1229, Washington University, St. Louis, MO 63130
Judith S. Weis (1996), Dept. of Biological Sciences, Rutgers University, Newark, NJ 07102
Annette W. Coleman (1997), Div. of Biology and Medicine, Brown University, Providence, RI 02912

Council Delegates

Gloria V. Callard (1995), Dept. of Biology, Boston University, 5 Cummington St., Boston, MA 02215
Susan Gottesman (1995), Bldg. 37, Rm. 4B03, Lab. of Molecular Biology, National Cancer Institute, Bethesda, MD 20892
Judith A. Lengyel (1995), Dept. of Biology, University of California, 405 Hilgard Ave., Los Angeles, CA 90024-1606
Nancy H. Marcus (1995), Dept. of Oceanography, Florida State University, Tallahassee, FL 32306
Ann G. Matthysse (1995), Dept. of Biology, Coker Hall, CB 3280, University of North Carolina, Chapel Hill, NC 27599
Diane K. Stoecker (1995), Horn Point Environmental Lab., University of Maryland System, CEES, P.O. Box 775, Cambridge, MD 21613
Zena Werb (1995), Lab. of Radiobiology and Environmental Health, LR-102, University of California, Box 0750, 3rd & Parnassus Aves., San Francisco, CA 94143-0750
Jeannette Yen (1995), Marine Sciences Research Center, State University of New York, Stony Brook, NY 11794-5000

Representatives of Affiliates

Janis B. Alcorn (1994; *Society for Economic Botany*), 4813 Morgan Dr., Chevy Chase, MD 20815
James G. Anderson (1995; *Society for Computer Simulation International*), 4141 Black Forest La., W. Lafayette, IN 47906

Peter C. Chabora (1996; *American Society of Naturalists*), Dept. of Biology, Queens College, CUNY, Flushing, NY 11367

Rufus Chaney (1996; *Society for Environmental Geochemistry and Health*), Soil-Microbial Systems Lab., AEQI-USDA-ARS, Rm. 108, Bldg. 318, BARC-East, Beltsville, MD 20705

John A. Chisler (1994; *West Virginia Academy of Science*), Div. of Science and Mathematics, Glenville State College, Glenville, WV 26351

David C. Culver (1994; *National Speleological Society*), Dept. of Biology, American University, 4400 Massachusetts Ave., N.W., Washington, DC 20016

Jeffrey Glenn Davis (1994; *Oak Ridge Associated Universities*), Oak Ridge Associated Universities, P.O. Box 117, Oak Ridge, TN 37831-0117

Vicki A. Funk (1996; *American Society of Plant Taxonomists*), Dept. of Botany, NHB 166, Smithsonian Institution, Washington, DC 20560

Robert S. Garofalo (1996; *New York Academy of Sciences*), Dept. of Anatomy and Cell Biology, SUNY Health Science Center, 450 Clarkson Ave., Box 5, Brooklyn, NY 11203

Patricia Gindhart (1995; *Anthropological Society of Washington*), Dept. of Anthropology, American University, 4400 Massachusetts Ave., N.W. Washington, DC 20016-8003

Patricia S. Goldman-Rakic (1994; *Society for Neuroscience*), Section of Neuroanatomy, Yale University School of Medicine, 333 Cedar St., New Haven, CT 06510

Bernard Goldstein (1995; *Society for the Scientific Study of Sex*), 111 Park Ave., San Carlos, CA 94070

Margaret Ann Goldstein (1996; *Microscopy Society of America*), Dept. of Medicine-CVS, Baylor College of Medicine, Houston, TX 77030-3498

Martin Hahn (1995; *Behavior Genetics Association*), Biology Dept., William Paterson College, Wayne, NJ 07470

Jay D. Hair (1996; *National Wildlife Federation*), National Wildlife Federation, 1400 16th St., N.W., Washington, DC 20036

Frederick W. Harrison (1995; *American Microscopical Society*), Dept. of Biology, Western Carolina University, Cullowhee, NC 28723

Richard Highton (1994; *Society for the Study of Evolution*), Dept. of Zoology, University of Maryland, College Park, MD 20742

Jerry Hirsch (1996; *Animal Behavior Society*), Dept. of Psychology, University of Illinois, 603 E. Daniel, Champaign, IL 61820

Rochelle Hirschhorn (1994; *American Society of Human Genetics*), Dept. of Medicine, New York University School of Medicine, 550 First Ave., New York, NY 10016

Mark Hite (1996; *Society of Toxicology*), Wyeth-Ayerst Research, 145 King of Prussia Rd., Radnor, PA 19087

David Ho (1996; *American Society of Plant Physiologists*), Dept. of Biology, Washington University, St. Louis, MO 63130

P. W. Hochachka (1996; *American Physiological Society*), Dept. of Zoology and Sports Medicine, University of British Columbia, 6270 University Blvd., #2354, Vancouver, B.C., Canada V6T 2A9

Harry E. Hodgdon (1994; *Wildlife Society*), Wildlife Society, 5410 Grosvenor La., Bethesda, MD 20814

Linda Hufnagel (1994; *Society of Protozoologists*), Dept. of Microbiology and Biophysics, University of Rhode Island, Kingston, RI 02881

Scott H. Hutchins (1996; *Entomological Society of America*), Dowelanco, P.O. Box 681428, Indianapolis, IN 46268-7428

John Karefa-Smart (1996; *World Population Society*), 4601 N. Park Ave., Chevy Chase, MD 20815

KyungMann Kim (1996; *Biometric Society, Eastern and Western North American Regions*), Dana-Farber Cancer Institute, 44 Binney St., Mayer 4, Boston, MA 02115

Robert S. Ledley (1994; *Computerized Medical Imaging Society*), National Biomedical Research Foundation, 3900 Reservoir Rd., N.W., Washington, DC 20007

David Lim (1995; *American Academy of Otolaryngology–Head and Neck Surgery*), Otolaryngology Research, Ohio State University, 456 W. 10th Ave., Rm. 4331, Columbus, OH 43210

Orie Loucks (1995; *American Institute of Biological Sciences*), Dept. of Zoology, Miami University, Oxford, OH 45056

Linda Mantel (1996; *Association for Women in Science*), Biology Dept., City College, Convent Ave. at 138th St., New York, NY 10031

Lawrence L. Master (1994; *The Nature Conservancy*), 201 Devonshire St., 5th Fl., Boston, MA 02110

Linda Maxson (1996; *American Society of Ichthyologists and Herpetologists*), Dept. of Biology, 208 Mueller Lab., Pennsylvania State University, University Park, PA 16802

Christos C. Mpelkas (1996; *Illuminating Engineering Society of North America*), 12 Mansfield St., Lynn, MA 01904

Steven C. Nelson (1995; *American Phytopathological Society*), American Phytopathological Society, 3340 Pilot Knob Rd., St. Paul, MN 55121

John S. Pearse (1995; *Western Society of Naturalists*), Institute of Marine Sciences, University of California, Santa Cruz, CA 95064

Bernard Phinney (1994; *Botanical Society of America*), Dept. of Biology, University of California, 405 Hilgard Ave., Los Angeles, CA 90024

William Presch (1994; *Society of Systematic Biologists*), Dept. of Biological Sciences, California State University, Fullerton, CA 92634

William D. Reese (1994; *American Bryological and Lichenological Society*), Dept. of Biology, University of Southwestern Louisiana, Lafayette, LA 70504-2451

Jerry C. Ritchie (1996; *Association of Southeastern Biologists*), USDA-ARS, Hydrology Lab., Bldg. 007, BARC-West, Beltsville, MD 20705

Theron S. Rumsey (1996; *American Society of Animal Science*), LPSI Ruminant Nutrition Lab., USDA-ARS, Bldg. 200, Rm. 124, BARC-East, Beltsville, MD 20705-2350

Edward I. Saiff (1995; *American Ornithologists Union*), Dept. of Biology, Ramapo College, Mahwah, NJ 07430

Sheldon J. Segal (1996; *Society for the Study of Social Biology*), The Population Council, 1 Dag Hammarskjold Plaza, 44th Fl., New York, NY 10017

Edwin C. Seim (1995; *Soil Science Society of America*), Dept. of Crop Science, California Polytechnic State University, San Luis Obispo, CA 93407

Patricia M. Shaffer (1996; *Sigma Delta Epsilon, Graduate Women in Science*), Dept. of Chemistry, University of San Diego, San Diego, CA 92110

Irving Shapiro (1994; *American Association for Dental Research*), School of Dental Medicine, Levy Bldg., University of Pennsylvania, 4001 Spruce St., Philadelphia, PA 19104

Lynda P. Shapiro (1996; *American Society of Limnology and Oceanography*), Oregon Institute of Marine Biology, University of Oregon, Charleston, OR 97420

Peter M. Sheehan (1994; *Paleontological Society*), Dept. of Geology, Milwaukee Public Museum, 800 W. Wells St., Milwaukee, WI 53233

L. Elliot Shubert (1994; *Phycological Society of America*), Dept. of Biology, University of North Dakota, Grand Forks, ND 58202

Gregory W. Siskind (1994; *Society for Experimental Biology and Medicine*), Cornell University Medical College, 1300 York Ave., Rm. A-131, New York, NY 10021

John M. Speer (1994; *Phi Sigma Biological Sciences Honor Society*), 207 Life Science, Botany Dept., Eastern Illinois University, Charleston, IL 61920

S. Phyllis Stearner (1995; *Foundation for Science and Disability*), 1141 Iroquois Dr., Apt. 114, Naperville, IL 60563-9376

Richard Storey (1994; *National Association of Biology Teachers*), Dept. of Biology, Colorado College, Colorado Springs, CO 80903

Frank Talamantes (1994; *Society for Advancement of Chicanos and Native Americans in Science*), Sinsheimer Labs., University of California, Santa Cruz, CA 95064

John W. Taylor (1994; *Mycological Society of America*), Dept. of Plant Biology, 111 Genetics and Plant Biology Bldg., University of California, Berkeley, CA 94720

Carroll P. Vance (1996; *Crop Science Society of America*), Dept. of Agronomy and Plant Science, University of Minnesota, 1509 Gortner, St. Paul, MN 55108

Arthur C. Washington (1994; *National Institute of Science*), Tennessee State University, Nashville, TN 37209

Lonnie L. Williamson (1994; *Wildlife Management Institute*), Wildlife Management Institute, 1101 14th St., N.W., Ste. 725, Washington, DC 20005

Leslie M. Yee (1996; *Society for Epidemiologic Research*), Proctor and Gamble Co., Ivorydale Technical Center, 5299 Spring Grove Ave., Cincinnati, OH 45217

CHEMISTRY

Officers

Chair: R. Stephen Berry, Dept. of Chemistry, University of Chicago, 5735 S. Ellis Ave., Chicago, IL 60637

Chair-Elect: Karen W. Morse, Provost, Utah State University, Logan, UT 84322-1435

Retiring Chair: Alvin L. Kwiram, Office of Research, AH-20, University of Washington, Seattle, WA 98195

Secretary: Robert W. Parry (1996), Dept. of Chemistry, University of Utah, Salt Lake City, UT 84112

Members-at-Large

Henry A. Bent (1994), Dept. of Chemistry, University of Pittsburgh, Pittsburgh, PA 15260
Ronald Breslow (1995), Dept. of Chemistry, 566 Chandler Labs., Columbia University, New York, NY 10027
J. Ivan Legg (1996), Provost, Memphis State University, Memphis, TN 38152
Edel Wasserman (1997), The du Pont Co., Central Research and Development, Experimental Station, P.O. Box 80328, Wilmington, DE 19880-0328

Council Delegates

Allen J. Bard (1995), Dept. of Chemistry, University of Texas, Austin, TX 78712
Kurt Mislow (1995), Dept. of Chemistry, Princeton University, Princeton, NJ 08544
John D. Roberts (1995), Crellin Lab. 164-30, California Institute of Technology, Pasadena, CA 91125
Nina M. Roscher (1995), Dept. of Chemistry, American University, 4400 Massachusetts Ave., N.W., Washington, DC 20016-8014

Representatives of Affiliates

Karl Boer (1994; *American Solar Energy Society*), Solar Knoll, 239 Bucktoe Hills Rd., Kennett Square, PA 19348
Ralph R. Booth (1994; *West Virginia Academy of Science*), Dept. of Chemistry, Davis & Elkins College, 100 Sycamore St., Elkins, WV 26241
B. Stephen Carpenter (1995; *American Nuclear Society*), Rm. B-113, Reactor Bldg., National Institute of Standards and Technology, Gaithersburg, MD 20899
C. Richard Cothern (1996; *Society for Environmental Geochemistry and Health*), USEPA, PM 222B, 401 M St., S.W., Washington, DC 20406
Brad T. Garber (1995: *American Industrial Hygiene Association*), 101 Hammonasset Meadows Rd., Madison, CT 06443
John Gavin (1995; *Foundation for Science and Disability*), 578 Tulip Poplar Crest, Carmel, IN 46032-1981
Fred M. Hawkridge (1995; *Electrochemical Society*), Dept. of Chemistry, Virginia Commonwealth University, Box 2006, Richmond, VA 23284-0001
Henry A. McGee, Jr. (1995; *American Institute of Chemical Engineers*), 706 Prince St., Apt. 9, Alexandria, VA 22314
Tyrone D. Mitchell (1995; *National Organization for the Professional Advancement of Black Chemists and Chemical Engineers*), Corning, Inc., R&D - SP DV. 0I 8, Corning, NY 14831
P. P. Nair (1995; *American Oil Chemists Society*), USDA, Lipid Nutrition Lab., Rm. 105, Bldg. 308, BARC-East, Beltsville, MD 20705
Robert Pecora (1995; *American Physical Society*), Dept. of Chemistry, Stanford University, Stanford, CA 94305
Jean'ne M. Shreeve (1995; *American Chemical Society*), University Research Office, University of Idaho, Moscow, ID 83843
Henry W. Strobel (1994; *American Society for Biochemistry and Molecular Biology*), University of Texas Medical School, P.O. Box 20708, Houston, TX 77225

Jeffrey D. Wolt (1995; *Soil Science Society of America*), 1801 Airfield La., Midland, MI 48642

DENTISTRY

Officers

Chair: John W. Stamm, CB 7450, Brauer Hall, University of North Carolina, Chapel Hill, NC 27599-7450
Chair-Elect: Marc W. Heft, Claude D. Pepper Center, JHMHSC, University of Florida, Box 100416, Gainesville, FL 32610
Retiring Chair: John S. Greenspan, Dept. of Stomatology, Rm. S-612, University of California, San Francisco, CA 94143-0422
Secretary: Thomas R. Dirksen (1995), School of Dentistry, Medical College of Georgia, Augusta, GA 30912-1000

Members-at-Large

Lawrence Tabak (1994), Dept. of Dental Research, University of Rochester, 601 Elmwood Ave., Box 611, Rochester, NY 14642
Lois K. Cohen (1995), National Institute of Dental Research, N.I.H., Westwood Bldg., Rm. 503, Bethesda, MD 20892
Barry R. Rifkin (1996), Div. of Basic Sciences, New York University College of Dentistry, 345 E. 24th St., New York, NY 10010
Brian H. Clarkson (1997), Dept. of Cariology and General Dentistry, School of Dentistry, University of Michigan, Ann Arbor, MI 48109-1078

Council Delegate

Kathleen L. Schroeder (1996), Dept. of Oral Pathology, School of Dentistry, Health Sciences Center, West Virginia University, Morgantown, WV 26506

Representatives of Affiliates

J. Henry Clarke (1996; *American Society of Clinical Hypnosis*), OHSU School of Dentistry, 611 S.W. Campus Dr., Portland, OR 97201
John J. Clarkson (1994; *American Association for Dental Research*), American Assn. for Dental Research, 1111 14th St., N.W., Ste. 1000, Washington, DC 20005
Donald B. Giddon (1994; *Society for Clinical and Experimental Hypnosis*), 170 Forest Ave., Newton, MA 02165-3012
Gordon H. Rovelstad (1995; *American Association of Dental Schools* and *American College of Dentists*), American College of Dentists, 7315 Wisconsin Ave., Ste. 352N, Bethesda, MD 20814
Chakwan Siew (1995; *American Dental Association*), Research Institute, American Dental Assn., 211 E. Chicago Ave., Chicago, IL 60611-2678

EDUCATION

Officers

Chair: Rodger W. Bybee, Biological Sciences Curriculum Study, 830 N. Tejon St., Ste. 405, Colorado Springs, CO 80903
Chair-Elect: Marvin Druger, Biology and Science Teaching, 103 Lyman Hall, Syracuse University, Syracuse, NY 13244
Retiring Chair: Madeleine J. Long, EHR, National Science Foundation, 1800 G St., N.W., Rm. 516, Washington, DC 20550
Secretary: Henry W. Heikkinen (1996), MAST Center, Ross Hall, University of Northern Colorado, Greeley, CO 80639

Members-at-Large

Marcia C. Linn (1994), Graduate School of Education, 4611 Tolman Hall, University of California, Berkeley, CA 94720
O. Roger Anderson (1995), Box 210, 525 W. 120th St., New York, NY 10027
Ronald D. Anderson (1996), CB 249, University of Colorado, Boulder, CO 80309
Raymond J. Hannapel (1997), National Science Foundation, 1800 G St., N.W., Washington, DC 20550

Council Delegate

Alice J. Moses (1996), ESIE, National Science Foundation, 1800 G St., N.W., Rm. 635A, Washington, DC 20550

Representatives of Affiliates

Efraim P. Armendariz (1995; *American Mathematical Society*), Dept. of Mathematics, University of Texas, Austin, TX 78712
J. Michael Armer (1996; *American Sociological Association*), Dept. of Sociology, Florida State University, Tallahassee, FL 32306
Gayle M. Ater (1994; *Sigma Pi Sigma*), Sigma Pi Sigma, 1825 Connecticut Ave., N.W., Ste. 213, Washington, DC 20009
Robert Beach (1996; *International Society for Educational Planning*), College of Education, 215 Ball Hall, Memphis State University, Memphis, TN 38152
Maurice Blaug (1995; *Association for Integrative Studies*), Hutchins School, Sonoma State University, Rohnert Park, CA 94928
Herbert K. Brunkhorst (1994; *National Science Teachers Association*), California State University, 5500 University Pkwy., San Bernardino, CA 92407
George Castro (1994; *Society for Advancement of Chicanos and Native Americans in Science*), IBM Almaden Research Center, 650 Harry Rd., San Jose, CA 95120
Walter Corson (1996; *World Population Society*), 1399 Orchard St., Alexandria, VA 22302
Wesley E. Cravey (1995; *Industrial Research Institute*), Nalco Chemical Co., P.O. Box 87, Sugar Land, TX 77487-0087

Lawrence Doolin (1995; *National Organization for the Professional Advancement of Black Chemists and Chemical Engineers*), 5320 E. Fall Creek Rd., Indianapolis, IN 46220

Isaac Eliezer (1995; *International Studies Association*), Office of the Dean, College of Arts and Sciences, Oakland University, Rochester, MI 48309

Darrel W. Fyffe (1994; *School Science and Mathematics Association*), 126 Life Sciences Bldg., Bowling Green State University, Bowling Green, OH 43403-0256

James L. Gaudino (1994; *Speech Communication Association*), Speech Communication Assn., 5105 Backlick Rd., Bldg. E, Annandale, VA 22003

Emmanuel K. Glakpe (1995; *American Nuclear Society*), Dept. of Mechanical Engineering, Howard University, Washington, DC 20059

Stanley I. Greenspan (1995; *American Psychoanalytic Association*), 7201 Glenbrook Rd., Bethesda, MD 20814

Helene N. Guttman (1995; *Scientists Center for Animal Welfare*), National Program Staff, USDA-ARS, Bldg. 002, Rm. 105, BARC-West, Beltsville, MD 20705

Paul A. Hanle (1996; *Maryland Academy of Sciences*), Maryland Science Center, 601 Light St., Baltimore, MD 21230

Maureen Hannley (1995; *American Academy of Otolaryngology–Head and Neck Surgery*), American Academy of Otolaryngology–Head and Neck Surgery, 1 Prince St., Alexandria, VA 22314

E. C. Keller, Jr. (1995; *Foundation for Science and Disability*), 236 Grand St., Morgantown, WV 26505

Jonathan Knight (1995; *American Association of University Professors*), American Assn. of University Professors, 1012 14th St., N.W., Ste. 500, Washington, DC 20005

Daniel Kunz (1995; *Junior Engineering Technical Society*), JETS, 1420 King St., Ste. 405, Alexandria, VA 22314

David A. Lanegran (1996; *Association of American Geographers*), Dept. of Geography, Carnegie Hall, Macalester College, St. Paul, MN 55105

LeRoy R. Lee (1994; *Wisconsin Academy of Sciences, Arts and Letters*), Wisconsin Academy of Sciences, Arts and Letters, 1922 University Ave., Madison, WI 53705-4099

Peter A. Lewis (1996; *Institute of Electrical and Electronics Engineers*), IEEE Service Center, P.O. Box 1331, Piscataway, NJ 08855-1331

Mary M. Lindquist (1996; *National Council of Teachers of Mathematics*), 14 7th St., Columbus, GA 31901

Theodore Lopushinsky (1994; *Society for College Science Teachers*), Rm. 100, N. Kedzie Lab., Michigan State University, E. Lansing, MI 48824

Jerrold William Maben (1996; *New York Academy of Sciences*), 33 Midbrook La., Greenwich, CT 06870-1427

Barbara Mandula (1996; *Association for Women in Science*), CNIE, 730 11th St., N.W., Ste. 200, Washington, DC 20001-4521

Glenn Markle (1994; *National Association for Research in Science Teaching*), National Assn. for Research in Science Teaching, University of Cincinnati, Cincinnati, OH 45221-0002

Celia L. Marshak (1996; *Sigma Delta Epsilon, Graduate Women in Science*), 430

Retaheim Way, La Jolla, CA 92037
Judith Moody (1995; *Association for Women Geoscientists*), J. B. Moody & Associates, 25 W. Washington St., Ste. 10, Athens, OH 45701-2447
Wayne A. Moyer (1994; *National Center for Science Education*), 1547 Scandia Cir., Reston, VA 22091
Joan Ferrini Mundy (1996; *Mathematical Association of America*), Dept. of Mathematics, University of New Hampshire, Durham, NH 03824
James V. O'Connor (1994; *National Association of Geology Teachers*), Dept. of Environmental Science, University of D.C., 4200 Connecticut Ave., N.W., B-44 R-203, Washington, DC 20008
John Padalino (1995; *American Nature Study Society*), Pocono Environmental Education Center, R.R. 2, Box 1010, Briscoe Mountain Rd., Dingmans Ferry, PA 18328
Ronald Rosier (1996; *Conference Board of the Mathematical Sciences*), Dept. of Mathematics, Georgetown University, Washington, DC 20057
Kenneth R. Roy (1995; *National Science Supervisors Association*), 82 Deepwood Dr., E. Hartford, CT 06118
Roy H. Saigo (1995; *American Institute of Biological Sciences*), Southwestern Louisiana University, P.O. Box 768, Hammond, LA 70404-0768
Brian B. Schwartz (1995; *American Physical Society*), American Physical Society, 335 E. 45th St., New York, NY 10017-3483
Cynthia M. Shewan (1996; *American Speech-Language-Hearing Association*), American Speech-Language-Hearing Assn., 10801 Rockville Pike, Rockville, MD 20852
Kendall N. Starkweather (1994; *International Technology Education Association*), International Technology Education Assn., 1914 Association Dr., Reston, VA 22091-1502
Juliana Texley (1994; *National Association of Biology Teachers*), Box 215, New Baltimore, MI 48047
Gerald F. Wheeler (1995; *American Association of Physics Teachers*), Science and Mathematics Resource Center, Montana State University, Bozeman, MT 59717
Paul C. Wohlmuth (1994; *International Society for the Systems Sciences*), School of Law, University of San Diego, Alcala Park, San Diego, CA 92110

ENGINEERING

Officers

Chair: Irene C. Peden, Div. of Electrical and Communications Systems, National Science Foundation, 1800 G St., N.W., Rm. 1151, Washington, DC 20550
Chair-Elect: Ernest S. Kuh, Dept. of ECS, University of California, Berkeley, CA 94720
Retiring Chair: Donald O. Pederson, 1436 Via Loma, Walnut Creek, CA 94598
Secretary: W. Edward Lear (1994), 5900 S.W. 35th Way, Gainesville, FL 32608

Members-at-Large

Alfred J. Eggers, Jr. (1994), RANN, Inc., 260 Sheridan Ave., Ste. 414, Palo Alto, CA 94306

Jose B. Cruz, Jr. (1995), College of Engineering, Ohio State University, 142 Hitchcock Hall, 2070 Neil Ave., Columbus, OH 43210-1275
Joseph Bordogna (1996), Directorate for Engineering, National Science Foundation, 1800 G St., N.W., Washington, DC 20550
Lawrence P. Grayson (1997), Office of Postsecondary Education, U.S. Dept. of Education, Washington, DC 20202-5151

Council Delegates

Nancy A. Da Silva (1994), Biochemical Engineering Program, University of California, Irvine, CA 92717
A. Richard Seebass (1994), Campus Box 422, University of Colorado, Boulder, CO 80309-0422

Representatives of Affiliates

Peter M. Bainum (1995; *American Astronautical Society*), 9804 Raleigh Tavern Ct., Bethesda, MD 20814
Teddy L. Barber (1995; *Foundation for Science and Disability*), 2833 El Camino Real, Las Cruces, NM 88005
Robert H. Bushnell (1994; *U.S. Metric Association*), 502 Ord Dr., Boulder, CO 80303
B. Leonard Carlson (1996; *Institute of Electrical and Electronics Engineers*), 516 W. Snoqualmie River Rd., S.E., Carnation, WA 98014
Halsey B. Chenoweth (1994; *Institute of Environmental Sciences*), Westinghouse Electric Corp., P.O. Box 746, MS 1655, Baltimore, MD 21203
Frank M. Coda (1996; *American Society of Heating, Refrigerating, and Air-Conditioning Engineers*), American Society of Heating, Refrigerating, and Air-Conditioning Engineers, 1791 Tullie Cir., N.E., Atlanta, GA 30329
Michael E. Fourney (1994; *Society for Experimental Mechanics*), Dept. of Civil Engineering, 4532 Boelter Hall, University of California, Los Angeles, CA 90024-1600
James D. Froula (1994; *Tau Beta Pi Association*), Tau Beta Pi Assn., P.O. Box 8840, University Station, Knoxville, TN 37996-4800
Richard E. Hallgren (1995; *American Meteorological Society*), American Meteorological Society, 45 Beacon St., Boston, MA 02108
Fred M. Hawkridge (1995; *Electrochemical Society*), Dept. of Chemistry, Virginia Commonwealth University, Box 2006, Richmond, VA 23284-0001
Dennis R. Heldman (1996; *American Society of Agricultural Engineers*), Food Science/Engineering Unit, 253 Agricultural Engineering Bldg., University of Missouri, Columbia, MO 65211
Joseph L. Holtshouser (1995; *American Industrial Hygiene Association*), Goodyear Tire & Rubber Co., 1144 E. Market St., Akron, OH 44316
John E. Kaufman (1996; *Illuminating Engineering Society of North America*), 1752 Newfield Ave., Stamford, CT 06903
Kazuhiko Kawamura (1994; *International Society for the Systems Sciences*), Center for Intelligent Systems, Vanderbilt University, Box 1674, Station B, Nashville, TN 37235

Samuel Kramer (1994; *National Society of Professional Engineers*), National Engineering Lab., National Institute of Standards and Technology, Bldg. 225, Rm. B-119, Gaithersburg, MD 20899

Daniel Kunz (1995; *Junior Engineering Technical Society*), JETS, 1420 King St., Ste. 405, Alexandria, VA 22314

Robert S. Ledley (1994, *Computerized Medical Imaging Society*; 1996, *Pattern Recognition Society*), National Biomedical Research Foundation, 3900 Reservoir Rd., N.W., Washington, DC 20007

Reuven R. Levary (1994; *Operations Research Society of America*), Dept. of Management and Decision Sciences, St. Louis University, St. Louis, MO 63108

Raymond P. Lutz (1995; *Institute of Industrial Engineers*), 10275 Hollow Way, Dallas, TX 75229

Gail H. Marcus (1995; *American Nuclear Society*), 8026 Cypress Grove La., Cabin John, MD 20818

Brijeshwar D. Mathur (1994; *Volunteers in Technical Assistance*), VITA, Inc., 1815 N. Lynn St., Ste. 200, Arlington, VA 22209-8438

Henry A. McGee, Jr. (1995; *American Institute of Chemical Engineers*), 706 Prince St., Apt. 9, Alexandria, VA 22314

Andrew H. Pettifor (1995, *Industrial Research Institute*; 1996, *American Institute of Aeronautics and Astronautics*), Rockwell International Science Center, P.O. Box 1085, Thousand Oaks, CA 91358

Alan L. Porter (1995; *International Association for Impact Assessment*), Dept. of Industrial and Systems Engineering, Georgia Institute of Technology, Atlanta, GA 30332

Robert J. Raudebaugh (1996; *ASM International*), 417 Stelle Ave., Plainfield, NJ 07060

Andrew Sage (1995; *American Society for Engineering Education*), School of Information Technology and Engineering, George Mason University, Fairfax, VA 22030

Murray Schwartz (1995; *American Ceramic Society*), Materials Technology Consulting, Inc., 30 Orchard Way N., Potomac, MD 20854

Bruce E. Seely (1994; *Society for the History of Technology*), Dept. of Social Sciences, Michigan Technological University, Houghton, MI 49931-1295

Bailus Walker, Jr. (1995; *American Public Health Association*), Health Sciences Center, University of Oklahoma, P.O. Box 26901, Oklahoma City, OK 73190

Leland J. Walker (1996; *American Society of Civil Engineers*), Northern Testing Labs., P.O. Box 7425, Great Falls, MT 59406-7425

GENERAL INTEREST IN SCIENCE AND ENGINEERING

Officers

Chair: Sharon M. Friedman, Dept. of Journalism and Communication, Lehigh University, 29 Trembley Dr., Bethlehem, PA 18015

Chair-Elect: Elizabeth S. Ivey, Provost's Office, Macalester College, 1600 Grand Ave., St. Paul, MN 55105

Retiring Chair: Sharon Dunwoody, School of Journalism and Mass Communication, University of Wisconsin, Madison, WI 53706

Secretary: James M. McCullough (1995), National Science Foundation, 1800 G St., N.W., Rm. 1230, Washington, DC 20550

Members-at-Large

Joann E. Rodgers (1994), Johns Hopkins Medical Institutions, Office of Public Affairs, 550 N. Broadway, 11th Fl., Baltimore, MD 21205-2011
Sharon M. Friedman (1995), Dept. of Journalism and Communication, Lehigh University, 29 Trembley Dr., Bethlehem, PA 18015
Sheila Grinell (1996), 86 Sherman Ave., Teaneck, NJ 07666
Mary Lynne Bird (1997), American Geographical Society, 156 Fifth Ave., Ste. 600, New York, NY 10010-7002

Council Delegate

Dael Wolfle (1996), Graduate School of Public Affairs, DC-13, University of Washington, Seattle, WA 98195

Representatives of Affiliates

John C. Bailar (1995; *Council of Biology Editors*), 468 N St., S.W., Washington, DC 20024
Elemer Bernath (1995; *Colorado-Wyoming Academy of Science*), 803 Ensign, Ft. Morgan, CO 80701
Mary Lynne Bird (1995; *American Geographical Society*), American Geographical Society, 156 Fifth Ave., Ste. 600, New York, NY 10010-7002
Robert H. Bushnell (1994; *U.S. Metric Association*), 502 Ord Dr., Boulder, CO 80303
Robert Fariel (1995; *National Science Supervisors Association*), National Science Supervisors Assn., P.O. Box AL, Amagansett, NY 11930
James H. Fribourgh (1995; *Arkansas Academy of Sciences*), University of Arkansas, 2801 S. University, Little Rock, AR 72204
James D. Froula (1994; *Tau Beta Pi Association*), Tau Beta Pi Assn., P.O. Box 8840, University Station, Knoxville, TN 37996-4800
Samuel Gingerich (1994; *South Dakota Academy of Science*), Dept. of Chemistry, Northern State University, Aberdeen, SD 57401
Patrick J. Gleason (1995; *Florida Academy of Sciences*), 1131 N. Palmway, Lake Worth, FL 33460
Paul G. Heltne (1995; *Chicago Academy of Sciences*), Chicago Academy of Sciences, 2001 N. Clark St., Chicago, IL 60614
James Henderson (1995; *Alabama Academy of Science*), Carver Research Lab., Tuskegee University, Tuskegee, AL 36088
William P. Hettinger, Jr. (1995, *Industrial Research Institute*; 1996, *Kentucky Academy of Science*), 203 Meadowlark Rd., Russell, KY 41169
Nancy Houk (1996; *Institute on Religion in an Age of Science*), Dept. of Astronomy, 1041 Dennison Bldg., University of Michigan, Ann Arbor, MI 48109-1090
David Hsi (1994; *New Mexico Academy of Science*), Agricultural Science Center, New Mexico State University, 1036 Miller St., S.W., Los Lunas, NM 87031
Craig Johnston (1995; *Montana Academy of Sciences*), School of Pharmacy and Allied

Health Sciences, PhP Bldg., Rm. 301, University of Montana, Missoula, MT 59812-1075
Donald M. Jordan (1994; *South Carolina Academy of Science*), College of Applied Professional Sciences, University of South Carolina, Columbia, SC 29208
Edward C. Keller (1994; *West Virginia Academy of Science*), Dept. of Biology, 237 Brooks Hall, West Virginia University, Morgantown, WV 26506
Gene Kritsky (1994; *Indiana Academy of Science*), Dept. of Biology, College of Mount St. Joseph, Mount St. Joseph, OH 45051
Daniel W. Kunz (1995; *Junior Engineering Technical Society*), JETS, 1420 King St., Ste. 405, Alexandria, VA 22314
Robert S. Ledley (1996; *Pattern Recognition Society*), National Biomedical Research Foundation, 3900 Reservoir Rd., N.W., Washington, DC 20007
LeRoy R. Lee (1994; *Wisconsin Academy of Sciences, Arts and Letters*), Wisconsin Academy of Sciences, Arts and Letters, 1922 University Ave., Madison, WI 53705-4099
Timothy Lynch (1994; *The Planetary Society*), The Planetary Society, 65 N. Catalina Ave., Pasadena, CA 91106
Clark Markell (1996; *North Dakota Academy of Science*), Div. of Science, Minot State University, 500 University Ave. W., Minot, ND 58702-5002
Henry A. McGee, Jr. (1995; *American Institute of Chemical Engineers*), 706 Prince St., Apt. 9, Alexandria, VA 22314
Blaine C. McKusick (1995; *Delaware Academy of Science*), 1212 Bruce Rd., Carrcroft, Wilmington, DE 19803
John McLeod (1995; *Society for Computer Simulation International*), 8484 La Jolla Shores Dr., La Jolla, CA 92037
Thomas O. McNearney (1994; *Academy of Science of St. Louis*), 5050 Oakland Ave., St. Louis, MO 63110
A. Lee Meyerson (1996; *New Jersey Academy of Science*), Beck Hall, Rm. 216, Box B, Rutgers University, Livingston Campus, Piscataway, NJ 08854
Edward N. Nelson (1995; *Oklahoma Academy of Science*), Dept. of Biology, Oral Roberts University, Tulsa, OK 74171
Claire Oswald (1996; *Nebraska Academy of Science*), Dept. of Biology, College of St. Mary, 1901 S. 72nd St., Omaha, NE 68124
Judith E. Parker (1996; *Minnesota Academy of Science*), 768 Summit Ave., St. Paul, MN 55105-3352
Dudley F. Peeler, Jr. (1996; *Mississippi Academy of Sciences*), Dept. of Neurosurgery, University of Mississippi Medical Center, 2500 N. State St., Jackson, MS 39216
Slobodan B. Petrovich (1995; *Association for Integrative Studies*), University of Maryland Baltimore County, 5401 Wilkens Ave., Baltimore, MD 21228-5398
Richard J. Raridon (1994; *Tennessee Academy of Science*), 111 Columbia Dr., Oak Ridge, TN 37830-7721
Paul E. Rider (1995; *Iowa Academy of Science*), 3538 MSH, University of Northern Iowa, Cedar Falls, IA 50614-0422
Albert Rothenberg (1996; *American Society for Aesthetics*), Austen Riggs Center, Inc., Stockbridge, MA 01262

George C. Shoffstall (1994; *Pennsylvania Academy of Science*), Pennsylvania Academy of Science, 502 Misty Dr., Ste. 1, Lancaster, PA 17603

James L. Smith (1996; *Missouri Academy of Science*), Dept. of Biology, Central Missouri State University, Warrensburg, MO 64093

George E. Stanton (1995; *Georgia Academy of Science*), Dept. of Biology, Science Hall 151A, Columbus College, Columbus, GA 31993-2399

Camm C. Swift (1995; *Southern California Academy of Sciences*), Southern California Academy of Sciences, 900 Exposition Blvd., Los Angeles, CA 90007

Ertle Thompson (1994; *Virginia Academy of Science*), Ruffner Hall, University of Virginia, 405 Emmet St., Charlottesville, VA 22903

Charles M. Vaughn (1994; *Ohio Academy of Science*), 6295 Devonshire Dr., Oxford, OH 45056

Jon M. Veigel (1994; *Oak Ridge Associated Universities*), Oak Ridge Associated Universities, P.O. Box 117, Oak Ridge, TN 37831-0117

GEOLOGY AND GEOGRAPHY

Officers

Chair: Jack Oliver, Dept. of Geological Sciences, 3120 Snee Hall, Cornell University, Ithaca, NY 14853

Chair-Elect: Richard S. Williams, Jr., U.S. Geological Survey, Quissett Campus, Woods Hole, MA 02543

Retiring Chair: Victor R. Baker, Dept. of Geosciences, University of Arizona, Tucson, AZ 85721

Secretary: Carroll Ann Hodges (1997), Branch of Resource Analysis, U.S. Geological Survey, 345 Middlefield Rd., MS 984, Menlo Park, CA 94025

Members-at-Large

David A. Ross (1994), Dept. of Geology and Geophysics, Woods Hole Oceanographic Institution, Woods Hole, MA 02543

Allison R. Palmer (1995), Institute for Cambrian Studies, 445 N. Cedarbrook Rd., Boulder, CO 80304

Charles B. Officer (1996), Dept. of Earth Sciences, Dartmouth College, Hanover, NH 03755-3571

James F. Hays (1997), Div. of Earth Sciences, National Science Foundation, 1800 G St., N.W., Washington, DC 20550

Council Delegate

George Rapp, Jr. (1995), Archaeometry Lab., University of Minnesota, Duluth, MN 55812-2496

Representatives of Affiliates

Keith M. Carr (1994; *The Nature Conservancy*), The Nature Conservancy, 1815 N. Lynn St., Arlington, VA 22209

Brian E. Davies (1996; *Society for Environmental Geochemistry and Health*), School of Environmental Science, University of Bradford, Bradford, W. Yorkshire BD7 1DP, England

Jessica Donovan (1995; *Association for Women Geoscientists*), Dames & Moore, 221 Main St., Ste. 600, San Francisco, CA 94105-1917

J. Thomas Dutro, Jr. (1994; *Geological Society of America* and *Paleontological Research Institution*), U.S. Geological Survey, National Museum of Natural History, Rm. E-308, Washington, DC 20560

Gerald M. Friedman (1994; *SEPM [Society for Sedimentary Geology]*), Northeastern Science Foundation, Rensselaer Center (CUNY), 15 Third St., P.O. Box 746, Troy, NY 12181

Louis Friedman (1994; *The Planetary Society*), The Planetary Society, 65 N. Catalina Ave., Pasadena, CA 91106

Thomas B. Griswold (1994; *American Institute of Professional Geologists*), Rhodes & Associates, Inc., P.O. Box 24080, Lexington, KY 40524-4080

John L. Hern (1994; *Society of Exploration Geophysicists*), J.L.H. Enterprises, 11767 Katy Frwy., Ste. 800, Houston, TX 77079

Vance T. Holliday (1995; *Soil Science Society of America*), Dept. of Geography, Science Hall, University of Wisconsin, Madison, WI 53706

Dana Isherwood (1995; *American Alpine Club*), Lawrence Livermore National Lab., P.O. Box 808, L-1, Livermore, CA 94550

Leonard Johnson (1994; *Seismological Society of America*), Div. of Earth Sciences, National Science Foundation, 1800 G St., N.W., Rm. 602, Washington, DC 20550

Marcus E. Milling (1995; *American Geological Institute*), American Geological Institute, 4220 King St., Alexandria, VA 22302

James K. Mitchell (1995; *American Geographical Society*), Dept. of Geography, Rutgers University, New Brunswick, NJ 08903

James V. O'Connor (1994; *National Association of Geology Teachers*), Dept. of Environmental Science, University of D.C., 4200 Connecticut Ave., N.W., B-44 R-203, Washington, DC 20008

John Pojeta, Jr. (1994; *Paleontological Society*), U.S. Geological Survey, National Center, MS 970, Reston, VA 22092

Edward C. Roy, Jr. (1994; *American Association of Petroleum Geologists*), Trinity University, 715 Stadium Dr., San Antonio, TX 78212

Mike Sfraga (1996; *Arctic Institute of North America*), 210 Signers Hall, University of Alaska, Fairbanks, AK 99775

John F. Shroder, Jr. (1996; *Association of American Geographers*), Dept. of Geography and Geology, University of Nebraska, Omaha, NE 68182-0199

James VanGundy (1994; *West Virginia Academy of Science*), Dept. of Biology and Environmental Sciences, Davis & Elkins College, 100 Sycamore St., Elkins, WV 26241

William B. White (1994; *National Speleological Society*), 210 Materials Research Lab., Pennsylvania State University, University Park, PA 16802

Warren Wood (1995; *Association of Ground Water Scientists and Engineers*), Water Resources Div., U.S. Geological Survey, National Center, MS 431, Reston, VA 22092

HISTORY AND PHILOSOPHY OF SCIENCE

Officers

Chair: Roger H. Stuewer, Tate Lab. of Physics, University of Minnesota, 116 Church St., S.E., Minneapolis, MN 55455
Chair-Elect: Jane Maienschein, Dept. of Philosophy, Arizona State University, Tempe, AZ 85287-2004
Retiring Chair: Michael Ruse, Dept. of Philosophy, University of Guelph, Guelph, Ont., Canada N1G 2W1
Secretary: Edward Manier (1997), 314 Decio Hall, University of Notre Dame, Notre Dame, IN 46556

Members-at-Large

John Beatty (1994), Dept. of Ecology, Evolution, and Behavior, University of Minnesota, Minneapolis, MN 55455
Merritt Roe Smith (1995), STS Program, Rm. E51-110, Massachusetts Institute of Technology, Cambridge, MA 02139
David L. Hull (1996), Dept. of Philosophy, Northwestern University, Evanston, IL 60208
Michael M. Sokal (1997), Dept. of Humanities, Worcester Polytechnic Institute, Worcester, MA 01609

Council Delegate

Jane Maienschein (1994), Dept. of Philosophy, Arizona State University, Tempe, AZ 85287-2004

Representatives of Affiliates

William P. Bechtel (1994; *Southern Society for Philosophy and Psychology*), Dept. of Philosophy, Georgia State University, University Plaza, Atlanta, GA 30303
Richard Burian (1996; *American Philosophical Association*), Center for the Study of Science in Society, 102 Price House, Virginia Polytechnic Institute and State University, Blacksburg, VA 24061-0247
C. West Churchman (1994; *International Society for the Systems Sciences*), P.O. Box 553, Bolinas, CA 94924
Edward Constant (1994; *Society for the History of Technology*), Dept. of History, Carnegie-Mellon University, Schenley Park, Pittsburgh, PA 15213
Mary Louise Gleason (1996; *History of Science Society*), 54 Riverside Dr., #3D, New York, NY 10024
Kenneth L. Manders (1996; *Association for Symbolic Logic*), Dept. of Philosophy, University of Pittsburgh, Pittsburgh, PA 15260
Edward Manier (1994; *Philosophy of Science Association*), 314 Decio Hall, University of Notre Dame, Notre Dame, IN 46556
Hart Nelsen (1994; *Society for the Scientific Study of Religion*), Dept. of Sociology, Pennsylvania State University, University Park, PA 16802
V. Frederick Rickey (1995; *American Mathematical Society*), Dept. of Mathematics

and Statistics, Bowling Green State University, Bowling Green, OH 43403-0221
Roger H. Stuewer (1994; *Sigma Pi Sigma*), Tate Lab. of Physics, University of Minnesota, 116 Church St., S.E., Minneapolis, MN 55455
Spencer Weart (1995; *American Institute of Physics*), American Institute of Physics, 335 E. 45th St., New York, NY 10017
Patricia Woolf (1995; *Society for Social Studies of Science*), 506 Quaker Rd., Princeton, NJ 08540

INDUSTRIAL SCIENCE

Officers

Chair: Jordan J. Baruch, Jordan Baruch Associates, 1200 18th St., N.W., Ste. 610, Washington, DC 20036
Chair-Elect: Theodore W. Schlie, CIMS, Rauch Business Center, Lehigh University, 621 Taylor St., Bethlehem, PA 18015-3117
Retiring Chair: Walter S. Baer, The RAND Corp., 1700 Main St., P.O. Box 2138, Santa Monica, CA 90406-2138
Secretary: Robert L. Stern (1996), 2000 P St., N.W., Ste. 608, Washington, DC 20036

Members-at-Large

Alden S. Bean (1994), CIMS, Rauch Business Center #37, Lehigh University, 621 Taylor St., Bethlehem, PA 18015
Joel D. Goldhar (1995), Stuart School of Business, Illinois Institute of Technology, 10 W. 31st St., Chicago, IL 60616
Paul E. Ritt (1996), 36 Sylvan La., Weston, MA 02193
James J. Solberg (1997), School of Industrial Engineering, Purdue University, W. Lafayette, IN 47907-1287

Council Delegate

Daniel Berg (1994), CII 5015, Rensselaer Polytechnic Institute, 110 8th St., Troy, NY 12180-3590

Representatives of Affiliates

Carl A. Bennett (1996; *American Statistical Association*), 11121 S.E. 59th St., Bellevue, WA 98006-2605
Alan D. Brailsford (1995; *American Institute of Physics*), Ford Motor Co., Scientific Research Lab., Rm. E 3172, Box 2053, Dearborn, MI 48121
James F. Frain (1995; *International Association for Impact Assessment*), 137 Cherry Valley Rd., Pittsburgh, PA 15221
Harold S. Goldberg (1996; *Institute of Electrical and Electronics Engineers*), 10 Alcott Rd., Lexington, MA 02173
Fred M. Hawkridge (1995; *Electrochemical Society*), Dept. of Chemistry, Virginia Commonwealth University, Box 2006, Richmond, VA 23284-0001
Joseph L. Holtshouser (1995; *American Industrial Hygiene Association*), Goodyear

Tire & Rubber Co., 1144 E. Market St., Akron, OH 44316
John McShefferty (1995; *Industrial Research Institute*), Gillette Research Institute, 401 Professional Dr., Gaithersburg, MD 20879
Sokrates Pantelides (1995; *American Physical Society*), IBM Watson Research Center, P.O. Box 218, Yorktown Heights, NY 10598
Bruce W. Schmeiser (1994; *Operations Research Society of America*), School of Industrial Engineering, Purdue University, W. Lafayette, IN 47907-1287
Wilford S. Stewart (1995; *National Organization for the Professional Advancement of Black Chemists and Chemical Engineers*), 6109 Carriage House La., Charlotte, NC 28226

INFORMATION, COMPUTING, AND COMMUNICATION

Officers

Chair: Bonnie C. Carroll, CENDI, P.O. Box 4219, Oak Ridge, TN 37831
Chair-Elect: Thelma Estrin, 201 Ocean Ave., #909P, Santa Monica, CA 90402
Retiring Chair: Toni Carbo Bearman, School of Library and Information Science, 505 LIS Bldg., University of Pittsburgh, 135 N. Bellefield St., Pittsburgh, PA 15260
Secretary: Elliot R. Siegel (1997), National Library of Medicine, 8600 Rockville Pike, Bethesda, MD 20894

Members-at-Large

Deborah Estrin (1994), Dept. of Computer Science, MC0781, University of Southern California, Los Angeles, CA 90089-0781
Peter G. Neumann (1995), SRI International, 333 Ravenswood Ave., EL-243, Menlo Park, CA 94025-3493
Maureen C. Kelly (1996), BIOSIS, 2100 Arch St., Philadelphia, PA 19103
Caroline M. Eastman (1997), Dept. of Computer Science, University of South Carolina, Columbia, SC 29208

Council Delegate

Susan L. Graham (1996), Computer Science Div., EECS, University of California, Berkeley, CA 94720

Representatives of Affiliates

John C. Bailar (1995; *Council of Biology Editors*), 468 N St., S.W., Washington, DC 20024
Bill Boland (1996; *New York Academy of Sciences*), New York Academy of Sciences, 2 E. 63rd St., New York, NY 10021
Robert L. Cox (1996; *International Communication Association*), 8140 Burnet Rd., Austin, TX 78758
Susan Crawford (1995; *Medical Library Association*), Medical Library, School of Medicine, Washington University, 660 S. Euclid Ave., St. Louis, MO 63110
Isaac Eliezer (1995; *International Studies Association*), Office of the Dean, College of Arts and Sciences, Oakland University, Rochester, MI 48309

Oscar N. Garcia (1996; *Institute of Electrical and Electronics Engineers*), 6308 Redwing Rd., Bethesda, MD 20817

Turkan K. Gardenier (1996; *American Statistical Association*), 1000 Salt Meadow La., McLean, VA 22101-2027

James L. Gaudino (1994; *Speech Communication Association*), Speech Communication Assn., 5105 Backlick Rd., Bldg. E, Annandale, VA 22003

Patricia S. Goldman-Rakic (1994; *Society for Neuroscience*), Section of Neuroanatomy, Yale University School of Medicine, 333 Cedar St., New Haven, CT 06510

Richard E. Hallgren (1995; *American Meteorological Society*), American Meteorological Society, 45 Beacon St., Boston, MA 02108

Frederick W. Harrison (1995; *American Microscopical Society*), Dept. of Biology, Western Carolina University, Cullowhee, NC 28723

T. C. Ingoldsby (1995; *American Institute of Physics*), Information Technology Branch, American Institute of Physics, 500 Sunnyside Blvd., Woodbury, NY 11797

Lou Joseph (1996; *American Medical Writers Association*), 1933 Locust St., Des Plaines, IL 60018

Zaven A. Karian (1996; *Mathematical Association of America*), Dept. of Mathematical Sciences, Denison University, Granville, OH 43023-1372

Helen A. Lawlor (1995; *American Chemical Society*), 276 Upper Gulph Rd., Radnor, PA 19807

Robert S. Ledley (1996; *Pattern Recognition Society*), National Biomedical Research Foundation, 3900 Reservoir Rd., N.W., Washington, DC 20007

Jay K. Lucker (1996; *American Library Association*), Director of Libraries, Massachusetts Institute of Technology, Cambridge, MA 02139

Clifford Lynch (1996; *American Society for Information Science*), Office of the President, University of California, 300 Lakeside Dr., 8th Fl., Oakland, CA 94612-3550

Carol L. Rogers (1994; *National Association of Science Writers*), 2908 Upton St., N.W., Washington, DC 20008

Ronald Rosier (1996; *Conference Board of the Mathematical Sciences*), Dept. of Mathematics, Georgetown University, Washington, DC 20057

Stephen R. Ruth (1996; *American Society for Cybernetics*), Expert Systems Development Center, George Mason University, Fairfax, VA 22030

Theodore D. Schultz (1995; *American Physical Society*), IBM Watson Research Center, Rm. 27-151, P.O. Box 218, Yorktown Heights, NY 10598

Chip G. Stockton (1995; *Society for Computer Simulation International*), Society for Computer Simulation International, P.O. Box 17900, San Diego, CA 92177-7900

Stanley C. Theobald (1996; *ASM International*), ASM International, Materials Park, OH 44073-0002

Melvin C. Thornton (1995; *American Mathematical Society*), Dept. of Mathematics, University of Nebraska, Lincoln, NE 68588-0323

Alfred Wohlpart (1994; *Oak Ridge Associated Universities*), Science Engineering Education Div., Oak Ridge Associated Universities, P.O. Box 117, Oak Ridge, TN 37831-0117

Marshall Yovits (1996; *Computing Research Association*), Dept. of Computer and Information Science, Indiana Purdue University, 723 W. Michigan St., SL 2260, Indianapolis, IN 46202-5132

LINGUISTICS AND LANGUAGE SCIENCES

Officers

Chair: Barbara C. Lust, Dept. of Human Development and Family Studies, Martha Van Rensselaer Hall, Cornell University, Ithaca, NY 14853
Chair-Elect: Richard S. Kayne, 470 West End Ave., New York, NY 10024
Retiring Chair: Kenneth L. Hale, Dept. of Linguistics, Massachusetts Institute of Technology, Cambridge, MA 02139
Secretary: Victoria A. Fromkin, Dept. of Linguistics, University of California, Los Angeles, CA 90024

Members-at-Large

Lise Menn (1994), 1625 Mariposa Ave., Boulder, CO 80302
David Pesetsky (1995), Dept. of Linguistics and Philosophy, Massachusetts Institute of Technology, 20D-219, Cambridge, MA 02139
Ilse Lehiste (1996), Dept. of Linguistics, 222 Oxley Hall, Ohio State University, 1712 Neil Ave. Columbus, OH 43210
Paul G. Chapin (1997), Linguistics Program, National Science Foundation, 1800 G St., NW, Rm. 320, Washington, DC 20550

Council Delegate

Arnold M. Zwicky (1996), Dept. of Linguistics, Ohio State University, 1712 Neil Ave., Columbus, OH 43210-1298

MATHEMATICS

Officers

Chair: R. L. Graham, AT&T Bell Labs., 600 Mountain Ave., Rm. 2T-102, Murray Hill, NJ 07974
Chair-Elect: Deborah Tepper Haimo, University of Missouri, 801 Natural Bridge Rd., St. Louis, MO 63121
Retiring Chair: Alice T. Schafer, Dept. of Mathematics, Marymount University, Arlington, VA 22207-4299
Secretary: Warren Page (1994), 30 Amberson Ave., Yonkers, NY 10705

Members-at-Large

Raymond O. Wells, Jr. (1994), Dept. of Mathematics, Rice University, Houston, TX 77251
Marian B. Pour-El (1995), School of Mathematics, University of Minnesota, Minneapolis, MN 55455
Richard Askey (1996), Dept. of Mathematics, University of Wisconsin, 480 Lincoln Dr., Madison, WI 53706
Fred Almgren (1997), 83 Riverside Dr., Princeton, NJ 08540

Council Delegate

Patricia J. Eberlein (1995), Dept. of Computer Science, 226 Bell Hall, State University of New York, Buffalo, NY 14260

Representatives of Affiliates

Jerry L. Bona (1996; *Mathematical Association of America*), Dept. of Mathematics, 215 McAllister Bldg., Pennsylvania State University, University Park, PA 16802

Margaret Butler (1995; *American Nuclear Society*), Argonne National Lab., 9700 S. Cass Ave., Bldg. 201, Argonne, IL 60439

John A. Dossey (1996; *National Council of Teachers of Mathematics*), R.R. 1, Box 165, Eureka, IL 61530

Raymond Johnson (1995; *American Mathematical Society*), Dept. of Mathematics, University of Maryland, College Park, MD 20742

Donald F. Kirwan (1994; *Sigma Pi Sigma*), Sigma Pi Sigma, 1825 Connecticut Ave., N.W., Ste. 213, Washington, DC 20009

Robert E. Levin (1996; *Illuminating Engineering Society of North America*), GTE Products Corp., 60 Boston St., Salem, MA 01970

Ronald Rosier (1996; *Conference Board of the Mathematical Sciences*), Dept. of Mathematics, Georgetown University, Washington, DC 20057

Christoph Witzgall (1994; *Operations Research Society of America*), Center for Computing and Applied Mathematics, National Institute of Standards and Technology, Gaithersburg, MD 20899

William Yslas Velez (1994; *Society for Advancement of Chicanos and Native Americans in Science*), Dept. of Mathematics, University of Arizona, Tucson, AZ 85721

MEDICAL SCIENCES

Officers

Chair: Richard J. Johns, Johns Hopkins Medical School, 1830 E. Monument St., Rm. 7300, Baltimore, MD 21205-2100

Chair-Elect: Charles C. J. Carpenter, Brown University, The Miriam Hospital, 164 Summit Ave., Providence, RI 02906

Retiring Chair: Edward N. Brandt, Jr., Rm. 359, CHB, University of Oklahoma, HSC, P.O. Box 26901, Oklahoma City, OK 73190

Secretary: Lewis H. Kuller (1994), GSPH, A527 Crabtree Hall, University of Pittsburgh, 130 DeSoto St., Pittsburgh, PA 15261

Members-at-Large

William R. Harlan (1994), National Institutes of Health, 9000 Rockville Pike, Bldg. 1, Rm. 260, Bethesda, MD 20892

William A. Blattner (1995), EPN/434, 6130 Executive Blvd., Rockville, MD 20892

Philip Fialkow (1996), School of Medicine, SC-59, University of Washington, 1959 N.E. Pacific, Seattle, WA 98195

Patricia A. Buffler (1997), 19 Earl Warren Hall, School of Public Health, University of California, Berkeley, CA 94720

Council Delegates

Abram S. Benenson (1994), Graduate School of Public Health, San Diego State University, San Diego, CA 92182

C. Gunnar Blomqvist (1994), Div. of Cardiology, H8.122, University of Texas Southwestern Medical Center, 5323 Harry Hines Blvd., Dallas, TX 75235-9034

Edwin C. Cadman (1994), Dept. of Internal Medicine, Yale University School of Medicine, 1072 LMP, 333 Cedar St., New Haven, CT 06510

Edward L. Kaplan (1994), Dept. of Pediatrics, Box 296, University of Minnesota Medical School, 420 Delaware St., S.E., Minneapolis, MN 55455

Howard E. Morgan (1994), Weis Center for Research, Geisinger Clinic, 100 N. Academy Ave., Danville, PA 17822-2601

Arno G. Motulsky (1994), Div. of Medical Genetics, RG-25, University of Washington, Seattle, WA 98195

David P. Rall (1994), 5302 Reno Rd., N.W., Washington, DC 20015

Representatives of Affiliates

Charles S. Adler (1996; *Association for Applied Psychophysiology and Biofeedback*), 955 Eudora St., #1605, Denver, CO 80220

J. Thomas August (1994; *American Society for Biochemistry and Molecular Biology*), Dept. of Pharmacology, Johns Hopkins University School of Medicine, 725 N. Wolf St., Baltimore, MD 21205

William R. Ayers (1994; *Computerized Medical Imaging Society*), Georgetown University Medical Center, 3900 Reservoir Rd., N.W., Washington, DC 20007

Kay A. Croker (1996; *American Society for Pharmacology and Experimental Therapeutics*), American Society for Pharmacology and Experimental Therapeutics, 9650 Rockville Pike, Bethesda, MD 20814-3995

Jeffrey Glenn Davis (1994; *Oak Ridge Associated Universities*), Oak Ridge Associated Universities, P.O. Box 117, Oak Ridge, TN 37831-0117

James A. Deye (1995; *American Association of Physicists in Medicine*), Dept. of Radiation Oncology, Fairfax Hospital, 3300 Gallows Rd., Falls Church, VA 22046

Everett H. Ellinwood, Jr., (1994; *Society of Biological Psychiatry*), Dept. of Psychiatry, Duke University Medical Center, Box 3870, Durham, NC 27710

Pennifer Erickson (1996; *American Statistical Association*), P.O. Box 2206, Wheaton, MD 20902

Dabney M. Ewin (1994; *Society for Clinical and Experimental Hypnosis*), 318 Baronne St., New Orleans, LA 70112-1606

Brad T. Garber (1995; *American Industrial Hygiene Association*), 101 Hammonasset Meadows Rd., Madison, CT 06443

Patricia S. Goldman-Rakic (1994; *Society for Neuroscience*), Section of Neuroanatomy, Yale University School of Medicine, 333 Cedar St., New Haven, CT 06510

John S. Greenspan (1994; *American Association for Dental Research*), Dept. of Stomatology, Rm. A-612, University of California, San Francisco, CA 94143-0422

Stanley I. Greenspan (1995; *American Psychoanalytic Association*), 7201 Glenbrook Rd., Bethesda, MD 20814

Margaret A. Hamburg (1996; *Society for the Study of Social Biology*), National Institute of Allergy and Infectious Diseases, N.I.H., Bldg. 31, Rm. 7A04, Bethesda, MD 20892

Rochelle Hirschhorn (1994; *American Society of Human Genetics*), Dept. of Medicine, New York University School of Medicine, 550 First Ave., New York, NY 10016

Mark Hite (1996; *Society of Toxicology*), Wyeth-Ayerst Research, 145 King of Prussia Rd., Radnor, PA 19087

Ariel Hollinshead (1996; *Sigma Delta Epsilon, Graduate Women in Science*), 3637 Van Ness St., N.W., Washington, DC 20008

Lou Joseph (1996; *American Medical Writers Association*), 1933 Locust St., Des Plaines, IL 60018

Kenneth Kendler (1995; *Behavior Genetics Association*), Dept. of Psychiatry, Medical College of Virginia, P.O. Box 710, Richmond, VA 23298-0710

Sherry Keramidas (1995; *American Physical Therapy Association*), American Physical Therapy Assn., 1111 N. Fairfax St., Alexandria, VA 22314

Jack L. Kostyo (1996; *American Physiological Society*), Dept. of Physiology, 7635 Medical Science II Bldg., Box 0622, University of Michigan Medical School, Ann Arbor, MI 48109

Stephen M. Krane (1994; *American College of Rheumatology*), Massachusetts General Hospital, Harvard Medical School, Fruit St., Boston, MA 02114

Robert Lavker (1994; *Society for Investigative Dermatology*), Duhring Labs., 235 CRB, University of Pennsylvania, Philadelphia, PA 19104-6142

J. Michael Marcum (1996; *American Society of Clinical Hypnosis*), 10 Riverside Dr., Binghamton, NY 13905

Arnauld Nicogossian (1995; *American Astronautical Society*), 9215 Bayard Pl., Fairfax, VA 22032

Grace L. Ostenso (1996; *American Dietetic Association*), 2871 Audubon Terr., N.W., Washington, DC 20008-2309

Jean Owen (1995; *American College of Radiology*), American College of Radiology, 1101 Market St., 14th Fl., Philadelphia, PA 19107

Gavril Pasternak (1996; *New York Academy of Sciences*), Dept. of Neurology, Memorial Sloan-Kettering Cancer Center, 1279 York Ave., New York, NY 10021

Harold Pincus (1995; *American Psychiatric Association*), American Psychiatric Assn., 1400 K St., N.W., Washington, DC 20005

William Richtsmeier (1995; *American Academy of Otolaryngology–Head and Neck Surgery*), 13027 Heilmanor Dr., Reistertown, MD 21136

Anthony Robbins (1995; *American Public Health Association*), School of Public Health, Boston University, 80 E. Concord St., Boston, MA 02118

Norman G. Sansing (1995; *Alpha Epsilon Delta*), 212F New College, University of Georgia, Athens, GA 30602

M. Roy Schwarz (1995; *American Medical Association*), American Medical Assn., 515 N. State St., Chicago, IL 60610

Cynthia M. Shewan (1996; *American Speech-Language-Hearing Association*), American Speech-Language-Hearing Assn., 10801 Rockville Pike, Rockville, MD 20852

Gregory W. Siskind (1994; *Society for Experimental Biology and Medicine*), Cornell University Medical College, 1300 York Ave., Rm. A-131, New York, NY 10021

Robert R. Smeby (1995; *Scientists Center for Animal Welfare*), Research Div., Cleveland Clinic Foundation, 9500 Euclid Ave., Cleveland, OH 44106

Herta Spencer-Laszlo (1996; *Society for Environmental Geochemistry and Health*), Veterans Administration, P.O. Box 35, Hines, IL 60141

F. William Sunderman, Jr. (1996; *Association of Clinical Scientists*), Dept. of Laboratory Medicine, University of Connecticut School of Medicine, 263 Farmington Ave., Farmington, CT 06032-9984

Jerome Wilson (1996; *Society for Epidemiologic Research*), Warner Lambert Co., 170 Tabor Rd., Morris Plains, NJ 07950

PHARMACEUTICAL SCIENCES

Officers

Chair: Gilbert S. Banker, College of Pharmacy, University of Iowa, Iowa City, IA 52242

Chair-Elect: Gary L. Grunewald, Dept. of Medicinal Chemistry, School of Pharmacy, 4060 Malott Hall, University of Kansas, Lawrence, KS 66045

Retiring Chair: John L. Neumeyer, Research Biochemicals, Inc., 1 Strathmore Rd., Natick, MA 01760

Secretary: Betty-ann Hoener (1995), School of Pharmacy, Box 0446, University of California, San Francisco, CA 94143-0446

Members-at-Large

Suzanne Frank Adair (1994), 4371 Edinburg Ct., Suisun City, CA 94585

Leslie Z. Benet (1995), Dept. of Pharmacy, Box 0446, University of California, San Francisco, CA 94143-0446

I. Glenn Sipes (1996), Dept. of Pharmacology and Toxicology, University of Arizona, Tucson, AZ 85721

Gordon L. Amidon (1997), 2012 College of Pharmacy, University of Michigan, Ann Arbor, MI 48109-1065

Council Delegate

Grant R. Wilkinson (1996), Dept. of Pharmacology, Vanderbilt University School of Medicine, Nashville, TN 37232-6600

Representatives of Affiliates

Albert A. Belmonte (1995; *American Pharmaceutical Association*), College of Pharmacy and Allied Health Professions, St. John's University, Jamaica, NY 11439

Patricia S. Goldman-Rakic (1994; *Society for Neuroscience*), Section of Neuroanatomy, Yale University School of Medicine, 333 Cedar St., New Haven, CT 06510

Mark Hite (1996; *Society of Toxicology*), Wyeth-Ayerst Research, 145 King of Prussia Rd., Radnor, PA 19087

Betty-ann Hoener (1995; *American Association of Pharmaceutical Scientists*), School of Pharmacy, Box 0446, University of California, San Francisco, CA 94143-0446

Steven M. Niemi (1995; *Scientists Center for Animal Welfare*), TSI Mason Labs., 57 Union St., Worcester, MA 01608

Joseph A. Oddis (1996; *American Society of Hospital Pharmacists*), American Society of Hospital Pharmacists, 7272 Wisconsin Ave., Bethesda, MD 20814

Carl E. Trinca (1995; *American Association of Colleges of Pharmacy*), American Assn. of Colleges of Pharmacy, 1426 Prince St., Alexandria, VA 22314

PHYSICS

Officers

Chair: Sam B. Treiman, Dept. of Physics, Princeton University, Princeton, NJ 08540

Chair-Elect: Raymond Orbach, Chancellor's Residence, 4171 Watkins Dr., Riverside, CA 92507

Retiring Chair: John P. Schiffer, Physics Div., Bldg. 203, Argonne National Lab., 9700 S. Cass Ave., Argonne, IL 60439-4843

Secretary: Rolf M. Sinclair (1996), National Science Foundation, 1800 G St., N.W., Washington, DC 20550

Members-at-Large

Wick Haxton (1994), Dept. of Physics, FM-15, University of Washington, Seattle, WA 98195

John W. Negele (1995), Rm. 6-308, Massachusetts Institute of Technology, Cambridge, MA 02139

Boris W. Batterman (1996), CHESS–Wilson Lab., Cornell University, Ithaca, NY 14853

Henry Ehrenreich (1997), Div. of Applied Sciences, Pierce Hall 205A, Harvard University, Cambridge, MA 02138

Council Delegates

Robert J. Birgeneau (1995), School of Science, Rm. 6-123, Massachusetts Institute of Technology, Cambridge, MA 02139

Jill C. Bonner (1995), Dept. of Physics, University of Rhode Island, Kingston, RI 02881

Representatives of Affiliates

George Abraham (1996; *Institute of Electrical and Electronics Engineers*), 3107 Westover Dr., S.E., Washington, DC 20020

Michael Buas (1994; *Computerized Medical Imaging Society*), Georgetown University Medical Center, 3900 Reservoir Rd., N.W., Washington DC 20007

Robert M. Fisher (1996; *Microscopy Society of America*), Dept. of Materials Science and Engineering, FB-10, University of Washington, Seattle, WA 98195

Edwin Goldin (1994; *Sigma Pi Sigma*), Sigma Pi Sigma, 1825 Connecticut Ave., N.W., Ste. 213, Washington, DC 20009

Richard E. Hallgren (1995; *American Meteorological Society*), American Meteorological Society, 45 Beacon St., Boston, MA 02108

Logan E. Hargrove (1995; *Acoustical Society of America*), Office of Naval Research, 800 N. Quincy St., Code 1112, Arlington, VA 22217-5000

Caroline Herzenberg (1996; *Association for Women in Science*), EAIS Div., Bldg. 900, Argonne National Lab., 9700 S. Cass Ave., Argonne, IL 60439

Clyde Jupiter (1995; *American Nuclear Society*), Jupiter Corp., Wheaton Plaza N., Ste. 503, 2730 University Blvd. W., Wheaton, MD 20902

David L. Nofziger (1995; *Soil Science Society of America*), Dept. of Agronomy, Oklahoma State University, Stillwater, OK 74078

Richard M. Obermann (1995; *American Astronautical Society*), 600 Alma St., S.E., Vienna, VA 22180

John S. Rigden (1995, *American Institute of Physics*; 1996, *Mathematical Association of America*), National Research Council, 2101 Constitution Ave., N.W., Washington, DC 20418-0001

Robert H. Romer (1995; *American Association of Physics Teachers*), Dept. of Physics, Amherst College, Amherst, MA 01002

Chih-Han Sah (1995; *American Mathematical Society*), Dept. of Mathematics, State University of New York, Stony Brook, NY 11794-3651

P. Kenneth Seidelmann (1996; *Institute of Navigation*), U.S. Naval Observatory, 34th & Massachusetts Ave., N.W., Washington, DC 20390

Robert Steele (1994; *National Institute of Science*), Dept. of Physics, Fort Valley State University, Fort Valley, GA 31030

Bruce W. Steiner (1994; *Optical Society of America*), Materials A-163, National Institute of Standards and Technology, Gaithersburg, MD 20899

N. Richard Werthamer (1995; *American Physical Society*), American Physical Society, 335 E. 45th St., New York, NY 10017-3483

Roland Winston (1994; *American Solar Energy Society*), Enrico Fermi Institute, 5640 S. Ellis Ave., Chicago, IL 60637

PSYCHOLOGY

Officers

Chair: Jerome Kagan, Dept. of Psychology, Harvard University, Cambridge, MA 02138

Chair-Elect: C. R. Gallistel, Dept. of Psychology, Franz Hall, University of California, Los Angeles, CA 90024-1563

Retiring Chair: William K. Estes, Dept. of Psychology, Harvard University, 33 Kirkland St., Cambridge, MA 02138

Secretary: William N. Dember (1994), Dept. of Psychology (ML 376), University of Cincinnati, Cincinnati, OH 45221-0376

Members-at-Large

Isidore Gormezano (1994), Dept. of Psychology, University of Iowa, Iowa City, IA 52242

James L. McGaugh (1995), Center for the Neurobiology of Learning and Memory, Bonney Center, University of California, Irvine, CA 92717-3800
Michael S. Gazzaniga (1996), Center for Neurobiology, University of California, Davis, CA 95616
Margaret Jean Intons-Peterson (1997), Dept. of Psychology, Indiana University, Bloomington, IN 47405-1301

Council Delegates

Byron A. Campbell (1994), Dept. of Psychology, Princeton University, Princeton, NJ 08544-1010
Larry R. Squire (1994), V.A. Medical Center (116A), 3350 La Jolla Village Dr., San Diego, CA 92161

Representatives of Affiliates

Henry L. Bennett (1994; *Society for Clinical and Experimental Hypnosis*), Dept. of Anesthesiology, UCD Medical Center, 2315 Stockton Blvd., Sacramento, CA 95817-2201
Gary G. Berntson (1996; *Society for Psychophysiological Research*), Dept. of Psychology, Ohio State University, 1885 Neil Ave., Columbus, OH 43210
Francine Butler (1996; *Association for Applied Psychophysiology and Biofeedback*), Assn. for Applied Psychophysiology and Biofeedback, 10200 W. 44th Ave., #304, Wheat Ridge, CO 80033
Ronald Cannella (1995; *Foundation for Science and Disability*), 28 Mill Rd., Manalapan, NJ 07726
Elizabeth D. Capaldi (1994; *Midwestern Psychological Association*), Dept. of Psychology, University of Florida, Gainesville, FL 32611-2065
Belinda L. Collins (1996; *Illuminating Engineering Society of North America*), National Institute of Standards and Technology, Bldg. 226, Rm. A313, Gaithersburg, MD 20899
William N. Dember (1996; *American Psychological Association*), Dept. of Psychology (ML 376), University of Cincinnati, Cincinnati, OH 45221-0376
Isaac Eliezer (1995; *International Studies Association*), Office of the Dean, College of Arts and Sciences, Oakland University, Rochester, MI 48309
James L. Gaudino (1994; *Speech Communication Association*), Speech Communication Assn., 5105 Backlick Rd., Bldg. E, Annandale, VA 22003
Patricia S. Goldman-Rakic (1994; *Society for Neuroscience*), Section of Neuroanatomy, Yale University School of Medicine, 333 Cedar St., New Haven, CT 06510
Stanley I. Greenspan (1995; *American Psychoanalytic Association*), 7201 Glenbrook Rd., Bethesda, MD 20814
Barbara Lust (1996; *Linguistic Society of America*), Dept. of Human Development, Van Rensselaer Hall, Cornell University, Ithaca, NY 14853
Gerald E. McClearn (1996; *Society for the Study of Social Biology*), Center for Developmental and Health Genetics, S-211, Henderson Bldg., Pennsylvania State University, University Park, PA 16802

James L. McGaugh (1994; *American Psychological Society*), Center for the Neurobiology of Learning and Memory, University of California, Irvine, CA 92717

Ralph E. McKinney (1996; *American Society of Clinical Hypnosis*), 5440 Balsam La. N., Plymouth, MN 55442

Daniel Offer (1995; *American Psychiatric Association*), Northwestern University Medical School, 303 E. Superior St., Chicago, IL 60611

James L. Pate (1994; *Southern Society for Philosophy and Psychology*), Dept. of Psychology, Georgia State University, University Plaza, Atlanta, GA 30303

David Rowe (1995; *Behavior Genetics Association*), Family and Consumer Resources, College of Agriculture, University of Arizona, Tucson, AZ 85721

Jeri Sechzer (1996; *Association for Women in Science*), Dept. of Psychology, Pace University, 1 Pace Plaza, New York, NY 10038-1502

Cynthia M. Shewan (1996; *American Speech-Language-Hearing Association*), American Speech-Language-Hearing Assn., 10801 Rockville Pike, Rockville, MD 20852

Thomas J. Tighe (1995; *Society for Research in Child Development*), Gulley Hall U-86, University of Connecticut, 352 Mansfield Rd., Storrs, CT 06268

Ethel Tobach (1996; *Eastern Psychological Association*), American Museum of Natural History, Central Park W. at 79th St., New York, NY 10024-5192

Arthur Washington (1994; *National Institute of Science*), Tennessee State University, Nashville, TN 37209

SOCIAL, ECONOMIC, AND POLITICAL SCIENCES

Officers

Chair: Garry D. Brewer, School of Natural Resources and Environment, Dana Bldg., University of Michigan, Ann Arbor, MI 48109-1115

Chair-Elect: Judith M. Tanur, 17 Longview Pl., Great Neck, NY 11021

Retiring Chair: David L. Featherman, Social Science Research Council, 605 Third Ave., New York, NY 10158

Secretary: William R. Freudenburg (1994), Dept. of Rural Sociology, University of Wisconsin, 1450 Linden Dr., Rm. 350, Madison, WI 53706

Members-at-large

Cora B. Marrett (1994), Social, Behavioral, and Economic Sciences, National Science Foundation, 1800 G St., N.W., Rm. 538, Washington, DC 20550

Larry L. Bumpass (1995), Dept. of Sociology, University of Wisconsin, 1180 Observatory Dr., Madison, WI 53711

Wendy Baldwin (1996), National Institute of Child Health and Human Development, Bldg. 31, Rm. 2A03, 9000 Rockville Pike, Bethesda, MD 20892

Marta Tienda (1997), Population Research Center, University of Chicago, 1155 E. 60th St., Rm. 167, Chicago, IL 60637-2799

Council Delegate

David L. Featherman (1994), Social Science Research Council, 605 Third Ave., New York, NY 10158

Representatives of Affiliates

Robert Beach (1996; *International Society for Educational Planning*), College of Education, 215 Ball Hall, Memphis State University, Memphis, TN 38152

Kenneth Boulding (1994; *International Society for the Systems Sciences*), P.O. Box 484, Boulder, CO 80309

Glenn Bugos (1994; *Society for the History of Technology*), Dept. of History, 3229 Dwinette Hall, University of California, Berkeley, CA 94720

Roland Chilton (1996; *American Society of Criminology*), Dept. of Sociology, University of Massachusetts, Amherst, MA 01003

Isaac Eliezer (1995; *International Studies Association*), Office of the Dean, College of Arts and Sciences, Oakland University, Rochester, MI 48309

James L. Gaudino (1994; *Speech Communication Association*), Speech Communication Assn., 5105 Backlick Rd., Bldg. E, Annandale, VA 22003

Scott D. Johnston (1994; *Pi Gamma Mu, International Honor Society in Social Science*), Dept. of Political Science, Hamline University, St. Paul, MN 55104

Roger E. Kasperson (1996; *Association of American Geographers*), Graduate School of Geography, Clark University, Worcester, MA 01610-1477

Ilsa A. Lottes (1995; *Society for the Scientific Study of Sex*), Dept. of Sociology, University of Maryland, 5401 Wilkens Ave., Baltimore, MD 21228

Phyllis Moen (1994; *American Sociological Association*), MVR Hall, Dept. of Sociology, Cornell University, Ithaca, NY 14853

Stuart S. Nagel (1995; *Policy Studies Organization*), 361 Lincoln Hall, University of Illinois, Urbana, IL 61801

Hart Nelsen (1994; *Society for the Scientific Study of Religion*), Dept. of Sociology, Pennsylvania State University, University Park, PA 16802

Valerie Oppenheimer (1994; *Population Association of America*), 10345 Strathmore Dr., Los Angeles, CA 90024

Jeffrey Passel (1996; *American Statistical Association*), 4146 N. 27th St., Arlington, VA 22207-5208

Matilda W. Riley (1996; *Society for the Study of Social Biology*), National Institute on Aging, N.I.H., Bldg. 31, Rm. 5C27, Bethesda, MD 20205

Frederick A. Rossini (1995; *International Association for Impact Assessment*), 13 Normandy Ct., N.E., Atlanta, GA 30324

Stephen R. Ruth (1996; *American Society for Cybernetics*), Expert Systems Development Center, George Mason University, Fairfax, VA 22030

Harvey M. Sapolsky (1995; *American Political Science Association*), Dept. of Political Science, Rm. E53-467, Massachusetts Institute of Technology, Cambridge, MA 02139

Gloria Scott (1996; *World Population Society*), 200 N. Pickett St., Ste. 1605, Alexandria, VA 22304

Carole L. Seyfrit (1995; *Rural Sociological Society*), Dept. of Sociology, P.O. Drawer C, Mississippi State University, Mississippi State, MS 39762

Thomas H. Tietenberg (1995; *American Economic Association*), Dept. of Economics, Colby College, Waterville, ME 04901

Jerome Wilson (1996; *Society for Epidemiologic Research*), Warner Lambert Co., 170 Tabor Rd., Morris Plains, NJ 07950

SOCIETAL IMPACTS OF SCIENCE AND ENGINEERING

Officers

Chair: Roberta Balstad Miller, Div. of SBER, National Science Foundation, 1800 G St., N.W., Washington, DC 20550
Chair-Elect: Rosemary Chalk, National Academy of Sciences, 2101 Constitution Ave., N.W., HA-172N, Washington, DC 20418
Retiring Chair: Glenn Paulson, Dept. of Environmental Engineering, Illinois Institute of Technology, Chicago, IL 60616
Secretary: Rachelle D. Hollander (1995), EVS, National Science Foundation, 1800 G St., N.W., Rm. 320, Washington, DC 20550

Members-at-Large

Don I. Phillips (1994), National Academy of Sciences, 2101 Constitution Ave., N.W., Washington, DC 20418
Jennifer Sue Bond (1995), Div. of Science Resources Studies, National Science Foundation, 1800 G St., N.W., Washington, DC 20550
J. William Futrell (1996), Environmental Law Institute, 1616 P St., N.W., Washington, DC 20036
Stephanie J. Bird (1997), Rm. 12-187, Massachusetts Institute of Technology, Cambridge, MA 02139

Council Delegate

Rachelle D. Hollander (1996), EVS, National Science Foundation, 1800 G St., N.W., Rm. 320, Washington, DC 20550

Representatives of Affiliates

Irving Adler (1996; *Vermont Academy of Arts and Sciences*), R.R. 1, Box 532, N. Bennington, VT 05257
John F. Ahearne (1994; *Sigma Xi, The Scientific Research Society*), P.O. Box 13975, Research Triangle Park, NC 27709
Ernst Benjamin (1995; *American Association of University Professors*), American Assn. of University Professors, 1012 14th St., N.W., Ste. 500, Washington, DC 20005
Alan J. Bennett (1995; *Industrial Research Institute*), Varian Associates, 611 Hansen Way, Palo Alto, CA 94303
J. Robert Berg (1995; *Kansas Academy of Science*), Dept. of Geology, Wichita State University, Wichita, KS 67208
Haven E. Bergeson (1996; *Utah Academy of Sciences, Arts and Letters*), Dept. of Physics, 201 JFB, University of Utah, Salt Lake City, UT 84112
Neil S. Berman (1995; *Arizona-Nevada Academy of Science*), Dept. of Chemical, Biological, and Materials Engineering, Arizona State University, Tempe, AZ 85287-6006
Stephanie J. Bird (1996; *Association for Women in Science*), Rm. 12-187, Massachusetts Institute of Technology, Cambridge, MA 02139

James P. Browder (1995; *American Academy of Otolaryngology–Head and Neck Surgery*), 733 Gimghoul Rd., Chapel Hill, NC 27514
William H. Clark (1995; *Idaho Academy of Science*), 6305 Kirkwood Rd., Boise, ID 83709
Eugene H. Cota-Robles (1994; *Society for Advancement of Chicanos and Native Americans in Science*), 2115 Pinehurst Ct., El Cerrito, CA 94530
Isaac Eliezer (1995; *International Studies Association*), Office of the Dean, College of Arts and Sciences, Oakland University, Rochester, MI 48309
Robert Fariel (1995; *National Science Supervisors Association*), National Science Supervisors Assn., P.O. Box AL, Amagansett, NY 11930
Virginia R. Ferris (1994; *Phi Beta Kappa*), Dept. of Entomology, Purdue University, W. Lafayette, IN 47907
Barbara Frase (1996; *Illinois State Academy of Science*), Dept. of Biology, Bradley University, Peoria, IL 61625
Maria C. Freire (1994; *Biophysical Society*), Graduate School, University of Maryland, Baltimore, MD 21201
Nord L. Gale (1996; *Society for Environmental Geochemistry and Health*), Life Sciences Dept., Rm. 105 Schrenk Hall, University of Missouri, Rolla, MO 65401
Michael I. Goldberg (1995; *American Society for Microbiology*), American Society for Microbiology, 1325 Massachusetts Ave., N.W., Washington, DC 20005
Bernice Goldsmith (1995; *International Association for Impact Assessment*), Concordia University, 1455 de Maisonneuve St. W., Montreal, P.Q., Canada H3G 1M8
Jay D. Hair (1996; *National Wildlife Federation*), National Wildlife Federation, 1400 16th St., N.W., Washington, DC 20036
William Hallahan (1994; *Rochester Academy of Science*), 6658 N. Avon Rd., Honeoye Falls, NY 14472
David Hsi (1994; *New Mexico Academy of Sciences*), Agricultural Science Center, New Mexico State University, 1036 Miller St., S.W., Los Lunas, NM 87031
Lou Joseph (1996; *American Medical Writers Association*), 1933 Locust St., Des Plaines, IL 60018
Solomon H. Katz (1996; *Institute on Religion in an Age of Science*), Dept. of Anthropology, University Museum, Rm. 341, University of Pennsylvania, Philadelphia, PA 19104
Daniel Kunz (1995; *Junior Engineering Technical Society*), JETS, 1420 King St., Ste. 405, Alexandria, VA 22314
Bruce V. Lewenstein (1996; *History of Science Society*), Dept. of Communication, 321 Kennedy Hall, Cornell University, Ithaca, NY 14853
Margaret M. Lobnitz (1996; *Sigma Delta Epsilon, Graduate Women in Science*), 4065 Woodman Ave., Sherman Oaks, CA 91423
Timothy Lynch (1994; *The Planetary Society*), The Planetary Society, 65 N. Catalina Ave., Pasadena, CA 91106
Wendy Mariner (1995; *American Public Health Association*), School of Public Health, Boston University, 80 E. Concord St., Boston, MA 02118-2394
Clark Markell (1996; *North Dakota Academy of Science*), Div. of Science, Minot State University, 500 University Ave. W., Minot, ND 58702-5002
Brijeshwar D. Mathur (1994; *Volunteers in Technical Assistance*), VITA, Inc., 1815 N. Lynn St., Ste. 200, Arlington, VA 22209-8438

A. Bradley McPherson (1996; *Louisiana Academy of Sciences*), Dept. of Biology, Box 41188, Centenary College, 2911 Centenary Blvd., Shreveport, LA 71104

Rodney W. Nichols (1996; *New York Academy of Sciences*), New York Academy of Sciences, 2 E. 63rd St., New York, NY 10021

Joel Orlen (1995; *American Academy of Arts and Sciences*), American Academy of Arts and Sciences, 136 Irving St., Cambridge, MA 02138

Slobodan B. Petrovich (1995; *Association for Integrative Studies*), University of Maryland Baltimore County, 5401 Wilkens Ave., Baltimore, MD 21228-5398

Nina M. Roscher (1995; *American Institute of Chemists*), Dept. of Chemistry, American University, 4400 Massachusetts Ave., N.W., Washington, DC 20016-8014

Ronald Rosier (1996; *Conference Board of the Mathematical Sciences*), Dept. of Mathematics, Georgetown University, Washington, DC 20057

Robert J. Rutman (1995; *U.S. Federation of Scholars and Scientists*), 3900 Ford Rd., Apt. PH-P, Philadelphia, PA 19131

Thomas A. Schroeder (1995; *Hawaiian Academy of Science*), Dept. of Meteorology, University of Hawaii, 2525 Correa Rd. (HIG 334), Honolulu, HI 96822

Richard C. Schwing (1995; *International Association for Impact Assessment*), Operating Sciences Dept., General Motors Labs., 30500 Mound Rd., Warren, MI 48090-9055

Kendall N. Starkweather (1994; *International Technology Education Association*), International Technology Education Assn., 1914 Association Dr., Reston, VA 22091-1502

Robert E. Stockhouse, II (1994; *Oregon Academy of Science*), Dept. of Biology, Pacific University, Forest Grove, OR 97116

Joseph Stoltman (1996; *Michigan Academy of Science, Arts and Letters*), Dept. of Geography, Western Michigan University, Kalamazoo, MI 49008

W. R. Stricklin (1996; *American Society of Animal Science*), Dept. of Animal Science, University of Maryland, College Park, MD 20742

J. Nathan Swift (1994; *National Association for Research in Science Teaching*), Education Dept., State University of New York, Oswego, NY 13126

Edward Taub (1996; *Association for Applied Psychophysiology and Biofeedback*), Dept. of Psychology, 201 Campbell Hall, University of Alabama, Birmingham, AL 35294

Jessica Utts (1994; *Parapsychological Association*), Div. of Statistics, University of California, Davis, CA 95616-8705

Armand B. Weiss (1994; *Washington Academy of Sciences*), 6516 Truman La., Falls Church, VA 22043

Sandra S. West (1994; *Texas Academy of Science*), Dept. of Biology, Southwest Texas State University, San Marcos, TX 78666-4616

R. Gary Williams (1995; *Rural Sociological Society*), 3316 N. 13th St., Arlington, VA 22201

Patricia Woolf (1995; *Society for Social Studies of Science*), 506 Quaker Rd., Princeton, NJ 08540

Leslie M. Yee (1996; *Society for Epidemiologic Research*), Proctor and Gamble Co., Ivorydale Technical Center, 5299 Spring Grove Ave., Cincinnati, OH 45217

STATISTICS

Officers

Chair: Daniel G. Horvitz, National Institute of Statistical Sciences, P.O. Box 14162, Research Triangle Park, NC 27709-4162

Chair-Elect: Mary Ellen Bock, 1399 Mathematical Sciences Bldg., Purdue University, W. Lafayette, IN 47907-1399

Retiring Chair: Donald B. Rubin, Dept. of Statistics, Harvard University, 1 Oxford St., Cambridge, MA 02138

Secretary: R. Clifton Bailey (1995), Health Standards and Quality, ME 2-D-2, Health Care Financing Administration, 6325 Security Blvd., Baltimore, MD 21207-5187

Members-at-Large

Gerald van Belle (1994), Dept. of Environmental Health, SC-32, School of Public Health and Community Medicine, University of Washington, Seattle, WA 98195

Peter J. Bickel (1995), Dept. of Statistics, University of California, Berkeley, CA 94720

Judith D. Goldberg (1996), Statistics and Data Management 60/203, Medical Research Div., Lederle Labs., Pearl River, NY 10965

Donald Guthrie (1997), Dept. of Psychiatry, University of California, Los Angeles, CA 90024-1759

Council Delegate

Nan M. Laird (1996), Dept. of Biostatistics, Harvard School of Public Health, 677 Huntington Ave., Boston, MA 02115

Representatives of Affiliates

Paul Glasserman (1994; *Operations Research Society of America*), 403 Uris Hall, Columbia University, New York, NY 10027

Mrudulla Gnanadesikan (1996; *Mathematical Association of America*), 425 Fairmount Ave., Chatham, NJ 07928-1369

Bruce Hoadley (1996; *American Statistical Association*), 16 Williamson Ct., Middletown, NJ 07748

Mary Grace Kovar (1995; *American Public Health Association*), National Center for Health Statistics, 6525 Belcrest Rd., Rm. 1120, Hyattsville, MD 20782

Robert Mare (1996; *American Sociological Association*), Dept. of Sociology, University of Wisconsin, 1180 Observatory Dr., Madison, WI 53706

Valerie Oppenheimer (1994; *Population Association of America*), 10345 Strathmore Dr., Los Angeles, CA 90024

Ronald Rosier (1996; *Conference Board of the Mathematical Sciences*), Dept. of Mathematics, Georgetown University, Washington, DC 20057

Leslie M. Yee (1996; *Society for Epidemiologic Research*), Proctor and Gamble Co., Ivorydale Technical Center, 5299 Spring Grove Ave., Cincinnati, OH 45217

Sections: Electorate Nominating Committees

The Electorate Nominating Committees are elected by the members of each section for the purpose of selecting candidates for various electorate and section offices.

AGRICULTURE

R. James Cook (1996), USDA-ARS, 367 Johnson Hall, Washington State University, Pullman, WA 99164

Harold D. Hafs (1995), Merck Research Labs., P.O. Box 2000, WBC 120, Rahway, NJ 07065

Betty Klepper (1996), USDA-ARS, P.O. Box 370, Pendleton, OR 97801

Margaret E. Smith (1994), Dept. of Plant Breeding, 252 Emerson Hall, Cornell University, Ithaca, NY 14853

Carroll P. Vance (1994), USDA-ARS, Dept. of Agronomy and Plant Genetics, 411 Borlaug Hall, University of Minnesota, St. Paul, MN 55108

Carol E. Windels (1995), 211 Agricultural Research Center, Northwest Experiment Station, University of Minnesota, Crookston, MN 56716

ANTHROPOLOGY

Cynthia M. Beall (1996), Dept. of Anthropology, 238 Mather Memorial Bldg., Case Western Reserve University, Cleveland, OH 44106-7125

Bernice A. Kaplan (1994), Dept. of Anthropology, Wayne State University, Detroit, MI 48202

William D. Lipe (1996), Dept. of Anthropology, Washington State University, Pullman, WA 99164-4910

David J. Meltzer (1994), Dept. of Anthropology, Southern Methodist University, Dallas, TX 75275

Sydel Silverman (1995), Wenner-Gren Foundation, 220 Fifth Ave., New York, NY 10001-7708

Phillip L. Walker (1995), Dept. of Anthropology, University of California, Santa Barbara, CA 93106

ASTRONOMY

Michael F. A'Hearn (1994), Dept. of Astronomy, University of Maryland, College Park, MD 20742-2421

John S. Gallagher, III (1995), Dept. of Astronomy, University of Wisconsin, 475 N. Charter St., Madison, WI 53706-1582

Note: Terms end on the last day of the Annual Meeting held in the years given in parentheses.

Karen B. Kwitter (1996), Dept. of Astronomy, Williams College, Williamstown, MA 01267

Susan Lea (1996), Dept. of Physics and Astronomy, San Francisco State University, San Francisco, CA 94132

Harvey Tananbaum (1994), Harvard-Smithsonian Center for Astrophysics, 60 Garden St., Bldg. B-426, Cambridge, MA 02138

Lee Anne Willson (1995), Dept. of Physics and Astronomy, Iowa State University, Ames, IA 50011

ATMOSPHERIC AND HYDROSPHERIC SCIENCES

Thomas P. Ackerman (1996), Dept. of Meteorology, 503 Walker Bldg., Pennsylvania State University, University Park, PA 16802

Eugene W. Bierly (1995), American Geophysical Union, 2000 Florida Ave., N.W., Washington, DC 20009

John C. Gille (1994), National Center for Atmospheric Research, P.O. Box 3000, Boulder, CO 80307-3000

Carol H. Pease (1995), NOAA Pacific Marine Environmental Lab., 7600 Sand Point Way, N.E., Seattle, WA 98115-0070

Joseph M. Prospero (1994), RSMAS, MAC, University of Miami, 4600 Rickenbacker Causeway, Miami, FL 33149-1098

Starley L. Thompson (1996), National Center for Atmospheric Research, P.O. Box 3000, Boulder, CO 80307-3000

BIOLOGICAL SCIENCES

May R. Berenbaum (1996), Dept. of Entomology, 320 Morrill Hall, University of Illinois, 505 S. Goodwin Ave., Urbana, IL 61801-3795

Ann Bucklin (1995), Dept. of Zoology, University of New Hampshire, Durham, NH 03824

Laurel R. Fox (1996), Ecology Program, National Science Foundation, 1800 G St., N.W., Washington, DC 20550

Virginia Walbot (1994), Dept. of Biological Sciences, Stanford University, Stanford, CA 94305-5020

Karen Wishner (1994), Graduate School of Oceanography, University of Rhode Island, S. Ferry Rd., Narragansett, RI 02882-1197

Paul H. Yancey (1995), Dept. of Biology, Whitman College, Walla Walla, WA 99362

CHEMISTRY

Edward M. Arnett (1995), 2529 Perkins Rd., Durham, NC 27706

Allen J. Bard (1996), Dept. of Chemistry, University of Texas, Austin, TX 78712

Jerome A. Berson (1994), Dept. of Chemistry, Yale University, 225 Prospect St., Box 6666, New Haven, CT 06511

Sharon K. Brauman (1996), Dept. of Chemistry, Stanford University, Stanford, CA 94305-5080

Helen M. Free (1995), Diagnostics Div., Miles, Inc., Box 70, Elkhart, IN 46515

Thomas J. Meyer (1994), Dept. of Chemistry, CB 3290, University of North Carolina, Chapel Hill, NC 27599-3290

DENTISTRY

Michael L. Barnett (1995), Warner-Lambert Co., 170 Tabor Rd., Morris Plains, NJ 07950

John J. Clarkson (1996), International/American Assns. for Dental Research, 1111 14th St., N.W., Ste. 1000, Washington, DC 20005

Roy S. Feldman (1995), Veterans Health Administration, Dental Service-160, University & Woodland Aves., Philadelphia, PA 19104

Frank G. Oppenheim (1994), School of Graduate Dentistry, Boston University Medical Center, 100 E. Newton St., Boston, MA 02118

Irving M. Shapiro (1996), Dept. of Biochemistry, School of Dental Medicine, University of Pennsylvania, 4001 Spruce St., Philadelphia, PA 19104-6003

John W. Stamm (1994), Brauer Hall, CB 7450, University of North Carolina, Chapel Hill, NC 27599-7450

EDUCATION

Mary M. Atwater (1995), Dept. of Science Education, 212 Aderhold Hall, University of Georgia, Athens, GA 30602

Diane M. Bunce (1996), Dept. of Chemistry, Catholic University of America, Washington, DC 20064

Alice J. Moses (1994), National Science Foundation, 1800 G St., N.W., Rm. 635A, Washington, DC 20550

Harold Pratt (1994), Middle School Life Science, Jefferson County Public Schools, 1829 Denver W. Dr., #27, Golden, CO 80401

Susan P. Speece (1996), Dept. of Biological Sciences, Anderson University, 1100 E. 5th St., Anderson, IN 46012

Emmett L. Wright (1995), 237 Bluemont Hall, C&I, Kansas State University, Manhattan, KS 66506-5301

ENGINEERING

Julia Abrahams (1996), Mathematical Sciences Div., Code 1111, Office of Naval Research, Arlington, VA 22217-5660

J. D. Achenbach (1994), Rm. 324, Catalysis Bldg., CTR QEFP, Northwestern University, 2137 N. Sheridan Rd., Evanston, IL 60208-3020

Alice M. Agogino (1996), Dept. of Mechanical Engineering, 5136 Etcheverry Hall, University of California, Berkeley, CA 94720

Frank S. Barnes (1995), Dept. of Electrical Engineering, University of Colorado, Boulder, CO 80303

John D. Kemper (1994), 1742 Midway Dr., Woodland, CA 95695

Terry E. Shoup (1995), School of Engineering, Santa Clara University, Santa Clara, CA 95053

GENERAL INTEREST IN SCIENCE AND ENGINEERING

Eduardo L. Feller (1995), Div. of International Programs, National Science Foundation, 1800 G St., N.W., V-501, Washington, DC 20550
Earle M. Holland (1994), University Communications, 1125 Kinnear Rd., Columbus, OH 43212
Jeanne V. Norberg (1995), Purdue News Service, 1132 ENAD, W. Lafayette, IN 47907-1132
Della M. Roy (1996), 217 MRL, Pennsylvania State University, University Park, PA 16802
John L. Safko (1996), Dept. of Physics and Astronomy, University of South Carolina, Columbia, SC 29208
Stanley Shapiro (1994), 8818 N. Kolmar, Skokie, IL 60076

GEOLOGY AND GEOGRAPHY

Pembroke J. Hart (1994), 3252 O St., N.W., Washington, DC 20007
John W. Hawley (1996), Bureau of Mines, New Mexico Tech, Socorro, NM 87801
James F. Miller (1996), Dept. of Geography, Geology, and Planning, Southwest Missouri State University, Springfield, MO 65804-0089
David K. Rea (1995), Dept. of Geological Sciences, University of Michigan, Ann Arbor, MI 48109-1063
Margaret N. Rees (1995), Dept. of Geoscience, University of Nevada, 4505 Maryland Pkwy., Las Vegas, NV 89154-4010
Robert E. Wall (1994), 14 Coburn Hall, University of Maine, Orono, ME 04469

HISTORY AND PHILOSOPHY OF SCIENCE

Michele L. Aldrich (1995), AAAS, 1333 H St., N.W., Washington, DC 20005
Stanley Goldberg (1995), 508 Third St., S.E., Washington, DC 20003
Anne Harrington (1996), Dept. of History of Science, Science Center 235, Harvard University, 1 Oxford St., Cambridge, MA 02138
Camille Limoges (1994), CREST, Universite du Quebec, Case Postale 8888, Succursale A, Montreal, Que., Canada H3C 3P8
Alan E. Shapiro (1994), School of Physics and Astronomy, University of Minnesota, 116 Church St., S.E., Minneapolis, MN 55455
Edith Dudley Sylla (1996), Dept. of History, Box 8108, North Carolina State University, Raleigh, NC 27695

INDUSTRIAL SCIENCE

Brian W. Burrows (1994), USG Research Center, 700 N. Highway 45, Libertyville, IL 60048
Ward J. Haas (1995), P.O. Box 644, Southport, CT 06490
Charles H. Kriebel (1996), Graduate School of Industrial Administration, Carnegie Mellon University, Pittsburgh, PA 15213

Lawrence M. Kushner (1995), 9528 Briar Glenn Way, Gaithersburg, MD 20879
Mary Ellen Mogee (1996), 212 Carrwood Rd., Great Falls, VA 22066
Theodore W. Schlie (1994), CIMS, Rauch Business Center, Lehigh University, 621 Taylor St., Bethlehem, PA 18015

INFORMATION, COMPUTING, AND COMMUNICATION

Joe Ann Clifton (1994), Box 850, HC 37, Kingman, AZ 86401
Deborah Estrin (1996), Dept. of Computer Science, MC0781, University of Southern California, Los Angeles, CA 90089-0781
Dorothy McGarry (1996), UCLA Physical Sciences and Technology Libraries, 8251 Boelter Hall, 405 Hilgard Ave., Los Angeles, CA 90024-1598
Jack Minker (1994), Dept. of Computer Science, University of Maryland, College Park, MD 20742
Shirley M. Radack (1995), Bldg. 225, Rm. B151, National Institute of Standards and Technology, Gaithersburg, MD 20899
Oliver R. Smoot (1995), Computer and Business Equipment Manufacturers Assn., 1250 I St., N.W., Ste. 200, Washington, DC 20005

LINGUISTICS AND LANGUAGE SCIENCES

Jill Carrier (1995), Dept. of Linguistics, Harvard University, Cambridge, MA 02138
Suzanne Flynn (1996), Massachusetts Institute of Technology, 14N321, Cambridge, MA 02139
Georgia M. Green (1994), Dept. of Linguistics, University of Illinois, Urbana, IL 61801
D. Terence Langendoen (1996), Dept. of Linguistics, University of Arizona, Douglass Bldg., Rm. 200E, Tucson, AZ 85721
Susan M. Steele (1994), Dept. of Linguistics, University of Arizona, Douglass 200E, Tucson, AZ 85721
Kenneth Wexler (1995), Massachusetts Institute of Technology, E10-020, 79 Amherst St., Cambridge, MA 02139

MATHEMATICS

R. Creighton Buck (1995), 3601 Sunset Dr., Madison, WI 53705
Barbara Lee Keyfitz (1996), Dept. of Mathematics, University of Houston, Houston, TX 77204-3476
Jill P. Mesirov (1995), Thinking Machines Corp., 245 First St., Cambridge, MA 02142
Mary Beth Ruskai (1994), 35 Pine St., Apt. A, Arlington, MA 02174
Paul J. Sally, Jr. (1996), Dept. of Mathematics, University of Chicago, Chicago, IL 60637
Jean E. Taylor (1994), Dept. of Mathematics, Rutgers University, New Brunswick, NJ 08903

MEDICAL SCIENCES

Peter S. Aronson (1996), Dept. of Medicine/Nephrology, Yale School of Medicine, New Haven, CT 06510

John G. Bartlett (1994), Div. of Infectious Diseases, Ross Research Bldg., Rm. 1159, 720 Rutland Ave., Baltimore, MD 21205

Bernard N. Fields (1995), Dept. of Microbiology and Molecular Genetics, Harvard Medical School, 200 Longwood Ave., Boston, MA 02115

Michael B. A. Oldstone (1996), Dept. of Neuropharmacology, IMM6, Scripps Research Institute, 10666 N. Torrey Pines Rd., La Jolla, CA 92037

David Schottenfeld (1994), Dept. of Epidemiology, SPHI, University of Michigan, 109 Observatory St., Ann Arbor, MI 48109-2029

Robert B. Wallace (1995), 2800 Steindler Bldg., University of Iowa, Iowa City, IA 52242

PHARMACEUTICAL SCIENCES

James D. McChesney (1996), Research Institute of Pharmaceutical Sciences, School of Pharmacy, University of Mississippi, University, MS 38677

Gerald T. Miwa (1995), Glaxo, Inc., 5 Moore Dr., Research Triangle Park, NC 27709

Paul R. Ortiz de Montellano (1995), School of Pharmacy, University of California, San Francisco, CA 94143-0446

Edward B. Roche (1994), College of Pharmacy, University of Nebraska Medical Center, 600 S. 42nd St., Omaha, NE 68198-6000

Elizabeth M. Topp (1996), Dept. of Pharmaceutical Chemistry, 3006 Malott Hall, University of Kansas, Lawrence, KS 66045-2504

Cheryl L. Zimmerman (1994), College of Pharmacy, University of Minnesota, 308 Harvard St., S.E., Minneapolis, MN 55455

PHYSICS

Jerome I. Friedman (1995), Rm. 24-512, Massachusetts Institute of Technology, Cambridge, MA 02139

Alan H. Guth (1995), Center for Theoretical Physics, Rm. 6-209, Massachusetts Institute of Technology, Cambridge, MA 02139

Michael L. Knotek (1994), Battelle Pacific Northwest Labs., P.O. Box 999, K1-48, Richland, WA 99352

Virginia Trimble (1996), Dept. of Physics, University of California, Irvine, CA 92717 (Jan.-June); Astronomy Program, Space Sciences Bldg., Rm. 1105, University of Maryland, College Park, MD 20742 (July-Dec.)

Daniel C. Tsui (1996), Dept. of Electrical Engineering, Princeton University, Princeton, NJ 08544

Vic Viola (1994), Cyclotron Facility, Indiana University, Bloomington, IN 47408

PSYCHOLOGY

Diana Deutsch (1994), Dept. of Psychology, University of California, San Diego, La Jolla, CA 92093

Frances K. Graham (1995), Dept. of Psychology, University of Delaware, Newark, DE 19716

James J. Jenkins (1994), Dept. of Psychology, BEH 339, University of South Florida, Tampa, FL 33620-8200
Mary C. Potter (1995), Dept. of Brain and Cognitive Sciences, Rm. E10-039, Massachusetts Institute of Technology, Cambridge, MA 02139
Irvin Rock (1996), Dept. of Psychology, University of California, Berkeley, CA 94720
John A. Swets (1996), BBN Labs., 10 Moulton St., Cambridge, MA 02138

SOCIAL, ECONOMIC, AND POLITICAL SCIENCES

Kenneth J. Arrow (1995), Dept. of Economics, Stanford University, Stanford, CA 94305-6072
Patricia A. Gwartney-Gibbs (1995), Dept. of Sociology, University of Oregon, Eugene, OR 97403-1291
Robert M. Hauser (1994), Dept. of Sociology, University of Wisconsin, 1180 Observatory Dr., Madison, WI 53706
Vernon W. Ruttan (1994), Dept. of Agricultural and Applied Economics, University of Minnesota, 1994 Buford Ave., 332 COB, St. Paul, MN 55108
Teresa A. Sullivan (1996), Main Bldg. 101, University of Texas, Austin, TX 78712
Michael S. Teitelbaum (1996), Alfred P. Sloan Foundation, 630 Fifth Ave., Ste. 2550, New York, NY 10111

SOCIETAL IMPACTS OF SCIENCE AND ENGINEERING

Janet Welsh Brown (1995), World Resources Institute, 1709 New York Ave., N.W., Ste. 700, Washington, DC 20006
Luther J. Carter (1996), 4522 Lowell St., N.W., Washington, DC 20016
Susan G. Hadden (1994), LBJ School of Public Affairs, University of Texas, Drawer Y, University Station, Austin, TX 78713-7450
Lois S. Peters (1996), 24 Deep Kill Rd., Troy, NY 12180
Alison G. Power (1995), Section of Ecology and Systematics, Corson Hall, Cornell University, Ithaca, NY 14853-2701
Paul Slovic (1994), Decision Research, 1201 Oak St., Eugene, OR 97401

STATISTICS

Nancy Flournoy (1996), Dept. of Mathematics and Statistics, American University, 4400 Massachusetts Ave., N.W., Washington, DC 20016-8050
Paul W. Holland (1994), Research Statistics Group, Educational Testing Service, MS 21, Princeton, NJ 08541
Thomas A. Louis (1996), Dept. of Biostatistics, University of Minnesota School of Public Health, 420 Delaware St., S.E., Box 197, Minneapolis, MN 55455
Juliet P. Shaffer (1995), Dept. of Statistics, University of California, Berkeley, CA 94720
Martin Tanner (1994), Div. of Biostatistics, University of Rochester Medical Center, P.O. Box 630, Rochester, NY 14642
Jessica Utts (1995), Div. of Statistics, University of California, Davis, CA 95616-8705

Regional Division Officers

ARCTIC DIVISION

President*: Arthur M. Pearson, Yukon Science Institute, Box 2799, Whitehorse, Yukon, Canada Y1A 5K4
President-Elect*: Rosa Meehan, U.S. Fish and Wildlife Service, 1011 E. Tudor Rd., Anchorage, AK 99503
Executive Secretary: Gunter Weller, Geophysical Institute, University of Alaska, Fairbanks, AK 99775-0800
*Through Sept. 30, 1993.

CARIBBEAN DIVISION

President*: Sergio Silva-Ruiz, Schering-Plough Products, Inc., P.O. Box 486, Manati, PR 00674-0486
President-Elect*: Ana Guadalupe, Dept. of Chemistry, University of Puerto Rico, Rio Piedras, PR 00931
Secretary-Treasurer: Lucy C. Gaspar, University High School, Box 23319, UPR Station, Rio Piedras, PR 00931-3319
*Through Dec. 1993.

PACIFIC DIVISION

President*: David Stoddart, Dept. of Geography, University of California, Berkeley, CA 94720
President-Elect*: Estella Leopold, Dept. of Botany, University of Washington, Seattle, WA 98195
Executive Director: Alan E. Leviton, California Academy of Sciences, Golden Gate Park, San Francisco, CA 94118
*Through June 24, 1994.

SOUTHWESTERN AND ROCKY MOUNTAIN DIVISION

President*: William B. Krantz, Dept. of Chemical Engineering, University of Colorado, Boulder, CO 80309-0424
President-Elect*: Donald R. Haragan, Dept. of Atmospheric Science, Texas Tech University, Lubbock, TX 79409
Executive Director: Donald J. Nash, Dept. of Biology, Colorado State University, Ft. Collins, CO 80523
*Through May 25, 1994.

II
Membership Organization Activities

Membership

The individual membership of the Association (defined in Article III of the Constitution [page 135]) mirrors the widening reach of the scientific and engineering enterprise. It includes many men and women who have achieved distinction both here and abroad for work at the leading edges of all fields of science and technology. Individual members are mainly research scientists, but many teachers and nonscientists concerned about the societal impacts of science and technology are represented in the work and activities of the Association. Among the latter are business executives, diplomats, journalists, ethicists, lawyers, managers and administrators, elected and career public servants, and members and supporters of public interest groups. All these, together with international members from developed and developing countries, combine to form a unique organization that reflects the versatility and significance of the sciences in human affairs.

Members receive the weekly journal *Science*. Members may purchase AAAS publications and audiotapes and register for the AAAS Annual Meeting at a discount. They may also participate in the group term life insurance plan, the AAAS Gold VISA card program, the Alamo car rental program and the AAAS travel program. In addition to full membership (currently $87/year), student, postdoctoral, emeritus, spouse, life, and international membership rates are available. Many professional scientists join the Association during their graduate student days; currently, dues for full-time students are $47/year.

All members are eligible for elective office, and the Board of Directors is elected from and by the general membership of the Association. The AAAS Council, which sets the general policies of the Association, is comprised primarily of individuals elected by the members through their section (electorate) elections. The Council acts on member resolutions and proposals received through responses to the "Call for Resolutions and Proposals" published in *Science* three months prior to each Annual Meeting. Members may also send proposals and resolutions to the AAAS executive officer at the AAAS address.

Fellows

A Fellow is "a Member whose efforts on behalf of the advancement of science or its applications are scientifically or socially distinguished." Nominations for Fellow are made by the section committee steering groups from the electorate membership rolls; by the Executive Officer; and by groups of three AAAS Fellows, provided that at least one of the three is not affiliated with the institution of the nominee. Election of Fellows takes place by mail in the fall of each year. For further information and a copy of the Fellow nomination form, write to the AAAS Executive Office, 1333 H St., N.W., Washington, DC 20005.

Section Membership

Members may select one of 23 sections of the Association as their primary electorate and focus for Association activities, but they may choose to vote in up to three sections in the annual elections. The sections are as follows: Agriculture; Anthropology; Astronomy; Atmospheric and Hydrospheric Sciences; Biological Sciences; Chemistry; Dentistry; Education; Engineering; General Interest in Science and Engineering; Geology and Geography; History and Philosophy of Science; Industrial Science; Information, Computing, and Communication; Linguistics and Language Sciences, Mathematics; Medical Sciences; Pharmaceutical Sciences; Physics; Psychology; Social, Economic, and Political Sciences; Societal Impacts of Science and Engineering; and Statistics.

Regional Divisions

Members may also participate in Association activities through the four geographic divisions of AAAS: Pacific, Southwestern and Rocky Mountain, Arctic, and Caribbean. Each division elects its own officers, holds annual meetings, and carries out other activities, including publication programs. Members wishing to participate in division activities should contact the executive officer of the division directly (see page 69).

Organization

The Association is directly governed by a 13-member **Board of Directors** (see page 5), with general policies set by an 81-member **Council** (see pages 6-9). All members are eligible to hold elective office and to participate in the annual election of the president-elect, members of the Board of Directors, and members of the Committee on Nominations. Members may petition to place names in nomination for any of these positions. Members vote to amend the Association's Constitution (see pages 135-138) and, while the Bylaws (see pages 139-152) are amendable only by the Council, members may propose amendments by petition.

SECTIONS

The 23 AAAS sections are determined by the Council and correspond to fields of interest of the membership (see page 141). Each member of the Association may enroll in from one to three sections (electorates), may vote for section candidates in each electorate in which he or she is enrolled, and is eligible for election to any position in those sections, except that no member may be elected to office in more than one section at a time (see Bylaw Article II, page 140). Section electorates vote in section elections for Council delegates, members of electorate nominating committees, section chairs-elect, and members-at-large of section committees.

Section affairs are managed by a section committee consisting of the section officers (retiring chair, chair, chair-elect, and secretary), four members-at-large, Council delegate(s) of the electorate, and one representative of each affiliate that is enrolled in the section. The chair-elect and one member-at-large are elected annually, the secretary is appointed for a four-year term by the section steering group, and the affiliate representatives are appointed by their organizations for three-year terms. (See pages 24-61 for current section committees and representatives.)

Under the direction of the section secretary, the section committee promotes the work of the Association in its own disciplinary area and proposes programs for presentation at AAAS Annual Meetings. Each section committee holds a business session at the AAAS Annual Meeting. Between meetings, a steering group consisting of the four section officers and four members-at-large is responsible for policy matters, for proposing nominees for election as Fellows, and, at four-year intervals, for appointing the section secretary.

ELECTORATES

Electorates consist of those members who are enrolled as voting members of the sections. They elect Council delegates, members of the electorate nominating committees, section chairs, and members-at-large of the section committees. Each electorate is represented on the Council by at least one delegate. When its membership reaches 3,000, an electorate elects a second delegate; when it reaches 6,000, a third delegate; and so on in increments of 3,000.

Membership by Electorate

Agriculture	1,589
Anthropology	954
Astronomy	828
Atmospheric and Hydrospheric Sciences	1,066
Biological Sciences	23,137
Chemistry	8,795
Dentistry	366
Education	1,053
Engineering	4,240
General Interest in Science and Engineering	808
Geology and Geography	2,320
History and Philosophy of Science	669
Industrial Science	364
Information, Computing, and Communication	1,768
Linguistics and Language Sciences	53
Mathematics	1,142
Medical Sciences	17,684
Pharmaceutical Sciences	1,066
Physics	4,671
Psychology	3,277
Social, Economic, and Political Sciences	1,428
Societal Impacts of Science and Engineering	1,274
Statistics	504
TOTAL	79,056

* As of 29 January 1993; members not enrolled in any electorate totaled 59,109.

REGIONAL DIVISIONS

Four regional divisions of the AAAS carry out the objectives of the Association in their own territories, electing their officers, holding their meetings, and managing their affairs independently. (See page 69 for a list of current officers.)

Arctic Division

Formerly the Alaska Division, the name of the Division was changed in 1982 to reflect the membership's growing interest in high latitudes outside of Alaska. Most of the 300 members of the Division reside in Alaska and Canada's Yukon and Northwest Territories, but any AAAS member who has an interest in the Arctic may join. The main activity of the Division is an annual meeting, which has taken place every year since 1950. Themes of recent meetings have included climate, fisheries, resources, and education. The Division has also been active in promoting the establishment of the U.S. Arctic Research Commission and a science advisor to the governor of Alaska, as well as pursuing contacts with scientists in the Russian Far East. The Division's president and executive committee of representatives are elected each year by mail ballot. The

Arctic and Pacific Divisions are planning to coorganize a major scientific conference in the Russian Far East in 1994 entitled "Bridges of Science Between North America and the Russian Far East: Past, Present and Future." The Russian coorganizer will be the Far East Branch of the Russian Academy of Sciences. For more information about this meeting or the Arctic Division, **contact** Dr. Gunter Weller, Executive Secretary, AAAS Arctic Division, Geophysical Institute, University of Alaska, Fairbanks, AK 99775-0800 (907-474-7371).

Caribbean Division

Established in 1985, the Caribbean Division is the newest AAAS Division. This Division has 490 AAAS members and covers the areas of Puerto Rico, Central America, islands of the Caribbean Basin, Venezuela, and southern Mexico. Each year, the Caribbean Division holds an Annual Meeting which is conducted in Spanish and generally held in Puerto Rico. The Caribbean Division's 1993 Annual Meeting will be held in San Juan in August or September in conjunction with the University of Puerto Rico NSF/EPSCoR Program. The Division has co-sponsored symposia on water, island ecology, new energy sources, and AIDS. In addition, the Caribbean Division works on various educational projects in conjunction with the Puerto Rico Science Teachers Association. New Division officers are elected every two years. For more information, **contact** Dr. Sergio A. Silva-Ruíz, President, AAAS Caribbean Division, 23 Mayagüez Street, Suite 10, Hato Rey, PR 00919.

Pacific Division

Founded in 1914, the Pacific Division of AAAS is the oldest and largest of the AAAS Divisions. Its membership includes nearly 32,000 AAAS members living in California, Hawaii, Idaho, western Montana, Nevada, Oregon, Utah, Washington, the Canadian provinces of British Columbia and Alberta, and all other countries bordering or lying within the Pacific Basin with the exception of mainland Mexico south to Panama. Twenty-five organizations are affiliated with the Division and several meet regularly with it at the Division's annual meeting, which is held in June. Professional scientists, teachers, and students are encouraged to participate in the meeting. The Division recognizes outstanding student research contributions by granting awards for excellence, including several that are endowed. The Division also sponsors field trips, workshops and conferences and has an active publications program. All members of AAAS are invited to participate in Pacific Division activities. To receive the Pacific Division's *Newsletter*, **contact** Dr. Alan E. Leviton, Executive Director, AAAS Pacific Division, California Academy of Sciences, Golden Gate Park, San Francisco, CA 94118 (415-752-1554).

Southwestern and Rocky Mountain Division

In 1920, scientists in the then rather isolated southwestern area decided to get together and form a Division of AAAS to further scientific communication in the area. The Southwestern Division came into being at that time. As the Rocky Mountain states were added later, the name was changed to include that region. The Division holds an annual meeting within its mid-continent geographic area; the 1993 meeting will be held in Albuquerque, New Mexico, 23-27 May. Future meetings are planned for Durango, Colorado (1994) and Norman, Oklahoma (1995). Students as well as professional scientists are encouraged to take part in the meetings and a variety of awards are provided for the best student presentations. An annual award to an outstanding community college teacher is also sponsored by the Division. To be added to the mailing list for Division publications, or for additional information, **contact** Dr. M. Michelle Balcomb, Executive Director, SWARM Division, AAAS, Colorado Mountain College, 215 Ninth Street, Glenwood Springs, CO 81601 (303-945-5516; fax 303-945-8902). After June 1, 1993, correspondence should be directed to Dr. Donald J. Nash, Department of Biology, Colorado State University, Fort Collins, CO 80523 (303-491-5481; fax 303-491-0649).

AFFILIATED ORGANIZATIONS

The 292 organizations affiliated with AAAS work with the Association on a variety of projects, including Annual Meeting symposia, international programs, annual analyses of the federal research and development budget, equal opportunity activities, and science education. There are 245 societies, 44 state and regional academies of science, and 3 city academies of science affiliated with AAAS. Each is wholly independent in its own special field or geographical area, but through their representation on section committees and in other programs, the affiliated organizations contribute much to the Association's strength.

Affiliated Organizations

Refer to individual section listings for affiliate sectional interests. Members of the AAAS Consortium of Affiliates for International Programs (see page 96) are indicated by an asterisk.

Acoustical Society of America
Alpha Epsilon Delta
American Academy of Arts and Sciences
American Academy of Forensic Sciences
American Academy of Otolaryngology Head and Neck Surgery
American Alpine Club
American Anthropological Association*
American Association for Dental Research
American Association of Anatomists
American Association of Blacks in Energy
American Association of Cereal Chemists*
American Association of Colleges of Pharmacy*
American Association of Dental Schools
American Association of Petroleum Geologists
American Association of Pharmaceutical Scientists
American Association of Physical Anthropologists
American Association of Physicists in Medicine
American Association of Physics Teachers*
American Association of University Professors*
American Astronautical Society*
American Astronomical Society

American Bryological and Lichenological Society
American Ceramic Society
American Chemical Society*
American College of Dentists*
American College of Radiology
American College of Rheumatology
American Dairy Science Association
American Dental Association
American Dietetic Association
American Economic Association*
American Ethnological Society*
American Fisheries Society*
American Geographical Society
American Geological Institute*
American Geophysical Union*
American Industrial Hygiene Association
American Institute of Aeronautics and Astronautics*
American Institute of Biological Sciences*
American Institute of Chemical Engineers*
American Institute of Chemists*
American Institute of Physics
American Institute of Professional Geologists
American Kinesiotherapy Association
American Library Association
American Mathematical Society*
American Medical Association*
American Medical Writers Association
American Meteorological Society*
American Microscopical Society
American Nature Study Society
American Nuclear Society*
American Oil Chemists Society*
American Ornithologists Union
American Pharmaceutical Association
American Philosophical Association
American Physical Society*
American Physical Therapy Association
American Physiological Society*
American Phytopathological Society*
American Political Science Association
American Psychiatric Association*
American Psychoanalytic Association
American Psychological Association*
American Psychological Society
American Public Health Association
American Society for Aesthetics
American Society for Biochemistry and Molecular Biology
American Society for Cybernetics*
American Society for Engineering Education*
American Society for Horticultural Science*
American Society for Information Science*
American Society for Microbiology*
American Society for Pharmacology and Experimental Therapeutics*
American Society for Photogrammetry and Remote Sensing
American Society of Agricultural Engineers*
American Society of Agronomy*
American Society of Animal Science*
American Society of Civil Engineers*
American Society of Clinical Hypnosis*
American Society of Criminology
American Society of Heating, Refrigerating, and Air-Conditioning Engineers
American Society of Hospital Pharmacists
American Society of Human Genetics
American Society of Ichthyologists and Herpetologists
American Society of Limnology and Oceanography
American Society of Mammalogists
American Society of Mechanical Engineers*
American Society of Naturalists
American Society of Plant Physiologists*
American Society of Plant Taxonomists
American Society of Zoologists*
American Sociological Association*
American Solar Energy Society
American Speech-Language-Hearing Association*
American Statistical Association*

AFFILIATED ORGANIZATIONS / 79

American Water Resources Association
Animal Behavior Society
Anthropological Society of Washington
Archaeological Institute of America
Arctic Institute of North America
ASM International*
Association for Applied Psychophysiology and Biofeedback
Association for Computing Machinery
Association for Integrative Studies
Association for Research on Nonprofit Organizations and Voluntary Action
Association for Symbolic Logic
Association for the Study of Man-Environment Relations
Association for Women Geoscientists
Association for Women in Mathematics
Association for Women in Science*
Association of American Geographers*
Association of Clinical Scientists
Association of Earth Science Editors
Association of Ground Water Scientists and Engineers
Association of Southeastern Biologists*
Astronomical Society of the Pacific

Behavior Genetics Association
Biometric Society, Eastern and Western North American Regions
Biophysical Society*
Botanical Society of America

Chi Beta Phi Scientific Fraternity
Computerized Medical Imaging Society
Computing Research Association
Conference Board of the Mathematical Sciences
Consortium on Peace Research, Education and Development
Cooper Ornithological Society
Council of Biology Editors
Crop Science Society of America

Eastern Psychological Association
Ecological Society of America*

The Electrochemical Society
Entomological Society of America*

Foundation for Science and Disability

Geochemical Society
Geological Society of America
Gerontological Society of America

History of Science Society*
Human Biology Council

Illuminating Engineering Society of North America
Industrial Research Institute
Institute of Electrical and Electronics Engineers*
Institute of Environmental Sciences
Institute of Food Technologists
Institute of Industrial Engineers
Institute of Management Sciences
Institute of Mathematical Statistics
Institute of Navigation*
Institute of Religion in an Age of Science
Instrument Society of America*
International Association for Impact Assessment*
International Communication Association*
International Society for Educational Planning
International Society for the Systems Sciences*
International Studies Association*
International Technology Education Association

Junior Engineering Technical Society

Linguistic Society of America*

Marine Technology Society*
Mathematical Association of America
Medical Library Association*
Microscopy Society of America*
Midwestern Psychological Association

Mycological Society of America

National Association for Research in Science Teaching*
National Association of Biology Teachers*
National Association of Geology Teachers
National Association of Science Writers
National Center for Science Education
National Council of Teachers of Mathematics*
National Institute of Science
National Marine Educators Association
National Organization for the Professional Advancement of Black Chemists and Chemical Engineers
National Science Supervisors Association
National Science Teachers Association*
National Society of Professional Engineers*
National Speleological Society*
National Wildlife Federation
Nature Conservancy

Oak Ridge Associated Universities
Operations Research Society of America*
Optical Society of America*
The Orton Dyslexia Society

Paleontological Research Institution
Paleontological Society
Parapsychological Association
Pattern Recognition Society
Phi Beta Kappa
Phi Sigma Biological Sciences Honor Society
Philosophy of Science Association
Phychological Society of America
Pi Gamma Mu, International Honor Society in Social Science
The Planetary Society*
Policy Studies Organization*
Population Association of America*
Poultry Science Association

Rural Sociological Society*

School Science and Mathematics Association*
Scientists Center for Animal Welfare
Seismological Society of America
SEPM (Society for Sedimentary Geology)
Sigma Delta Epsilon, Graduate Women in Science*
Sigma Pi Sigma
Sigma Xi, The Scientific Research Society*
Society for Advancement of Chicanos and Native Americans in Science
Society for American Archaeology
Society for Applied Anthropology*
Society for Clinical and Experimental Hypnosis
Society for College Science Teachers
Society for Computer Simulation International
Society for Economic Botany*
Society for Environmental Geochemistry and Health
Society for Epidemiologic Research
Society for Experimental Biology and Medicine
Society for Experimental Mechanics
Society for Industrial and Applied Mathematics
Society for Investigative Dermatology
Society for Neuroscience
Society for Psychophysiological Research
Society for Research in Child Development*
Society for Social Studies of Science*
Society for the History of Technology
Society for the Scientific Study of Religion*
Society for the Scientific Study of Sex
Society for the Study of Evolution
Society for the Study of Social Biology
Society of American Foresters*
Society of Biological Psychiatry
Society of Exploration Geophysicists*
Society of Protozoologists
Society of Systematic Biologists
Society of Toxicology
Soil and Water Conservation Society

Soil Science Society of America*
Southern Society for Philosophy and Psychology*
Speech Communication Association*

Tau Beta Pi Association

U.S. Federation of Scholars and Scientists

U.S. Metric Association*
Volunteers in Technical Assistance*

Western Society of Naturalists
Wildlife Management Institute
The Wildlife Society*
World Population Society

Affiliated Academies of Science

Academy of Science of St. Louis
Alabama Academy of Science
Arizona-Nevada Academy of Science
Arkansas Academy of Science

California Academy of Sciences
Chicago Academy of Sciences
Colorado–Wyoming Academy of Science

Delaware Academy of Science

Florida Academy of Sciences

Georgia Academy of Science

Hawaiian Academy of Science

Idaho Academy of Science
Illinois State Academy of Science
Indiana Academy of Science
Iowa Academy of Science

Kansas Academy of Science
Kentucky Academy of Science

Louisiana Academy of Sciences

Maryland Academy of Sciences
Michigan Academy of Science, Arts and Letters
Minnesota Academy of Science
Mississippi Academy of Sciences
Missouri Academy of Science

Montana Academy of Sciences
Nebraska Academy of Sciences
New Jersey Academy of Science
New Mexico Academy of Science
New York Academy of Sciences
North Carolina Academy of Science
North Dakota Academy of Sciences
Northwest Scientific Association

Ohio Academy of Science
Oklahoma Academy of Science
Oregon Academy of Science

Pennsylvania Academy of Science

Rochester Academy of Science

South Carolina Academy of Science
South Dakota Academy of Science
Southern California Academy of Sciences

Tennessee Academy of Science
Texas Academy of Science

Utah Academy of Sciences, Arts and Letters

Vermont Academy of Arts and Sciences
Virginia Academy of Science

Washington Academy of Sciences
West Virginia Academy of Science
Wisconsin Academy of Sciences, Arts and Letters

National Association of Academies of Science (NAAS)

With few exceptions, the state, regional, and municipal academies of science have come into existence since the founding of AAAS in 1848, a number of them with Association assistance. Forty-seven academies of science are now AAAS affiliates. In 1927, an Academy Conference was authorized by the AAAS Council to serve as a standing committee on relations among the affiliated academies and between all of them and the AAAS. The Academy Conference met for the first time at the 1928 New York meeting of the Association, and it has presented programs at subsequent Annual Meetings. In 1969 the name of the Academy Conference was changed to Association of Academies of Science, and in 1979 to the National Association of Academies of Science. The governing body of the NAAS is composed of two representatives from each affiliated academy. The NAAS names two delegates to the AAAS Council.

Financed by allotments from its constituent organizations, the NAAS assists new academies to organize and keeps all the academies informed of each other's activities. The NAAS has a strong interest in the work of junior and collegiate academies of science and in encouraging young people interested in science. The academies are eligible to receive funds annually from AAAS for support of research by high school students (see page 108). At each AAAS Annual Meeting, representatives of state junior academies present scientific papers. In addition, each year each affiliated academy may nominate for a one-year membership in AAAS two secondary school and two college students. These honorary junior memberships in AAAS include a subscription to *Science*. For further information, write to the AAAS Directorate for Education and Human Resources Programs at the AAAS address.

The 1993 NAAS officers are given below. For more information about NAAS activities, contact Lynn Elfner, NAAS Archivist, at the address below or call 614-424-6045.

President: David Hsi, NMSU Agricultural Science Center, 1036 Miller St., SW, Los Lunas, NM 87031

Past President: Don Jordan, South Carolina Academy, Mathematical Sciences, College of Applied Professional Sciences, University of South Carolina, Columbia, SC 29208

Secretary: Dudley F. Peeler, Mississippi Academy, University Medical Center, Jackson, MS 39216

Treasurer: Ollin Drennan, Missouri Academy, NE Missouri State University, Kirksville, MO 63501

Archivist: Lynn E. Elfner, Ohio Academy, 1500 West Third Ave., Ste. 223, Columbus, OH 43212-2817

Director, AJAS: Gloria J. Takahashi, Southern California Academy, 900 Exposition Blvd., Los Angeles, CA 90007

Council Delegates: Claire Oswald, Nebraska Academy, 1901 S. 72nd St., Omaha, NE 68124 and Richard J. Raridon, Tennessee Academy, 111 Columbia Dr., Oak Ridge, TN 37830

AAAS Representatives: Jerry Bell and Judy Kass, 1333 H St., NW, Washington, DC 20005

PARTICIPATING ORGANIZATIONS

Commission on Professionals in Science and Technology (CPST)

The Commission on Professionals in Science and Technology was established in 1953 as the Scientific Manpower Commission, a nonprofit corporation concerned with the recruitment, training, and utilization of scientific and technical personnel. In addition to the leading scientific societies that founded the Commission in 1953 and were its only members until 1986, the expanded, renamed Commission is now open to membership by other engineering and related professional societies, corporations, institutions, and individuals who share its concerns.

The Commission continues its program of collecting, analyzing, publishing, and disseminating statistical data and other information to aid the development of the country's scientific, engineering, and technological resources for the benefit and welfare of all its people. Publications include a monthly periodical, *Scientific, Engineering, Technical Manpower Comments,* summarizing current developments affecting the recruitment, training, and utilization of scientific, engineering, and technical professionals, as well as numerous special reports on various aspects of the technological community. In addition, the Commission serves as an S/E personnel resource office for the AAAS and cooperates closely with it, particularly in the areas of education, communications, and science and society programs. Beginning with the Eleventh Edition in 1993, a major Commission publication, *Professional Women and Minorities: A Total Human Resource Data Compendium,* will be published and distributed by the AAAS Press. Inquiries should be addressed to Betty Vetter, Executive Director, or to Eleanor Babco, Associate Director, CPST, 1500 Massachusetts Ave., NW, Ste. 831, Washington, DC 20005.

Gordon Research Conferences: "Frontiers of Science"

The Gordon Research Conferences (formerly the Gibson Island Research Conferences) were established on a permanent basis under AAAS auspices in 1938 and renamed in 1948 to honor their founder, Neil E. Gordon of Johns Hopkins University and later Wayne State University. In 1955 the AAAS Council named the Gordon Research Conferences a participating organization of the Association.

The purpose of the conferences is to stimulate research in universities, research foundations, and industrial laboratories. Each conference includes scheduled lectures and group discussions; afternoons are free for recreation, reading, and informal discussion. From two in the first year (1931), the conferences have expanded to more than 130 at present. Both summer and winter conferences are held, and attendance at each is usually limited to 115. To promote open and free discussion, information presented at the conferences is not to be cited or published without the specific authorization of the individual making the contribution, whether in formal presentation or in discussion. Scientific reports are not published as emanating from the conferences.

The complete programs are published in *Science,* in March for the summer conferences and in October for the winter conferences. Requests to attend the conferences should be addressed to Alexander M. Cruickshank, Director, Gordon

Research Conferences, Gordon Research Center, University of Rhode Island, Kingston, RI 02881 (401-783-4011).

Illinois Science Lecture Association

The Illinois Science Lecture Association (ISLA) is a nonprofit corporation chartered in 1971 to bring to Chicago a series of Christmas-time lectures in the physical and biological sciences. The ISLA is funded by corporate contributions, and, in 1978, became a participating organization of the Association.

The lectures, given in the Main Auditorium at the Museum of Science and Industry in Chicago, present a distinguished science program primarily for high-ability, scienceoriented secondary school seniors and juniors and their science teachers. Modeled after the London Christmas Lectures founded by Michael Faraday, the lectures are given by eminent scholars who have helped shape their fields of science and who seek to convey a sense of the adventure in science. The program includes informal discussions with guest speakers and faculty members from the University of Chicago, University of Illinois at Chicago, Northwestern University, and Illinois Institute of Technology, which co-sponsor the lectures.

In addition to the Christmas lectures, ISLA presents a one-day series, Spring Science Lectures, specifically for junior high school students. Inquiries about either lecture series should be addressed to the Illinois Science Lecture Association, c/o Museum of Science and Industry, Education Dept., 57th St. and Lake Shore Dr., Chicago, IL 60637-2093.

Activities

PUBLICATIONS

The American Association for the Advancement of Science is the publisher of the journal *Science, Science Books & Films,* and *The Online Journal of Current Clinical Trials,* and of books covering a wide range of topics. AAAS also publishes a variety of newsletters, reports, and other materials that derive from and support the goals of the Directorates and Offices of the Association.

Science

The AAAS publishes the weekly journal *Science,* received by all members as a part of their membership and available by subscription to nonmembers and institutions. Founded by Thomas A. Edison and since 1900 the official journal of the AAAS, *Science* serves as a forum for the presentation and discussion of important issues related to the advancement of science, including minority or conflicting points of view. Emphasis is on material pertaining to the interactions among science, technology, government, and society. The peer-reviewed section of the journal presents cutting-edge research of either interdisciplinary interest or unusual significance to the specialist. Topical coverage reflects the range of AAAS's interest across the physical, biological, and social sciences.

Science is the highest circulation journal for the interdisciplinary scientific community, a major resource for journalists who report scientific research in lay terms, and one of the most frequently cited scientific magazines in the world. Major sections of the magazine are This Week in *Science,* News and Comment, Research News, Articles and Research Articles, and Research Reports. The AAAS Communications Office works with journalists throughout the year to ensure strong media coverage of the fast-breaking research, significant advances and new frontiers reported in those sections. *Science* also includes book reviews, editorial and letters sections, policy forums, perspectives, software reviews, and information about new instrumentation. A monthly section, Inside AAAS, informs members about activities of the Association.

Science is indexed in one general periodical index, the *Reader's Guide to Periodical Literature,* and in more than three dozen specialized indexing and abstracting journals. Volumes of *Science* on microfilm and microfiche are available back to Vol. 1, 1880, from University Microfilms International, 300 N. Zeeb Rd., Ann Arbor, MI 48106. *Science* is also included in the Information Access full-text database and is available through the Magazine ASAP file on Dialog. In addition, the full text is available on CD-ROM through UMI's General Periodicals on disc database. The AAAS also publishes the annual *Guide to Biotechnology Products and Instruments.*

Reprints of material published in *Science* are available from stock for classroom or other use. Collections of reprints by subject will be prepared on demand. For information and prices, write to the Reprints Sales Office at the AAAS address.

Science Books & Films (SB&F)

Written for librarians, media specialists, curriculum supervisors, science teachers, and others responsible for recommending or purchasing science materials, *Science Books & Films* is published nine times per year. It provides critical reviews of the scientific accuracy and presentation of print, audiovisual, and electronic resources intended for use in science, technology, and mathematics education. Subscription cost $40 for one year (nine issues). For subscription information, call 202-326-6446; members interested in reviewing materials should contact Maria Sosa, 202-326-6453.

The Online Journal of Current Clinical Trials

This peer reviewed electronic journal was launched in July 1992. CCT is published by AAAS in a joint venture with OCLC Online Computer Library Center, Inc., of Dublin, Ohio. The journal, which is available by subscription, publishes research reports, reviews, metaanalyses, editorials, news, and letters describing or commenting on trials of therapies, procedures, and other interventions relevant to patient care in all fields and subspecialties of medicine. The journal is indexed and abstracted in BIOSIS. Reprints of articles are available. For information about subscriptions or reprints, write to the Publications Office at the AAAS address, or call 202-326-6446.

Books

The publications catalog describes current titles. In addition to books developed by the Directorates, the list includes: titles in a symposium series developed from meetings sponsored by outside organizations as well as by the AAAS; collected reprints from *Science*; and titles from AAAS Press. The Press, established in 1992, works with series editors to acquire titles in areas that are central to the AAAS mission: science, technology, and mathematics education; science, technology, and health policy; and environment and global change. To obtain a copy of the current publications catalog, write to the Publications Office at the AAAS address, or call 202-326-6446.

Reports/Newsletters/Directories

In addition to books listed in the publications catalog, each Directorate develops materials in aid of program activities. For information on those publications, write to individual Directorates at the AAAS address, or call the telephone or facsimile numbers listed on page 2 of this handbook.

AAAS Meetings on Audiocassette

Many symposia from AAAS annual meetings and colloquia are available on audiocassette. For details regarding meetings in 1993 or later, contact National Recording Services, 15385 S. 159 Hwy., Olathe, KS 66062 (913-780-3307; fax: 913-780-5091). For details regarding meetings prior to 1993 contact Mobiltape Company, Inc., 25061 W. Avenue Stanford, Suite 70, Valencia, CA 91355 (805-295-0504; fax: 805-295-8474).

AAAS Selected Symposia Series

Between 1978 and 1990, 113 books based on symposia from AAAS annual meetings were published for AAAS by Westview Press, Dept. AAAS, 5500 Central Avenue, Boulder, CO 80301. Contact Westview directly for a catalog of available titles.

MEETINGS

Annual National Meeting

Since its first meeting in Philadelphia on 20 September 1848, the AAAS has held 158 national meetings, skipping some years due to wars or other emergencies and holding both a summer and winter meeting in others (see page 121). Annual meetings are now held in February.

Because the Association's membership is drawn from professionals in all major areas of science, Annual Meeting sessions focus on interdisciplinary treatments of topics. Of particular interest are problems involving the interactions of science and society and science education. The scientific meeting format includes symposia, topical and plenary lectures, specialized seminars, poster presentations, and an exhibition of science societies, publishers, and equipment manufacturers, as well as activities tailored for local middle and high school students.

Each year posters submitted by undergraduate and graduate students for consideration for Student Poster Awards are judged by a panel of distinguished AAAS fellows and other scientists. Awards in three categories—life, physical, and social sciences—include cash and a complimentary one-year AAAS membership.

Association meetings are fully accessible to scientists with disabilities; the AAAS is a pioneer in the conduct of barrier-free meetings.

Broad media coverage of annual meetings helps further the public understanding of science. The AAAS Communications Office sends program information to hundreds of reporters in advance of the meeting and provides texts of many papers in the newsroom, which is busier than that at any other U.S. scientific meeting. The meetings are extensively covered nationally and internationally by reporters from major newspapers, news magazines, wire services, and radio and television organizations.

Future Annual Meetings

1994 San Francisco (18-23 February)
1995 Atlanta (16-21 February)
1996 Baltimore (8-13 February)
1997 Seattle (13-18 February)
1998 Philadelphia (12-17 February) sesquicentennial celebration!

Suggestions for Annual Meeting sessions (symposia, seminars, lectures) should be sent to the AAAS Meetings Office, 1333 H Street, NW, Washington, DC 20005, before April 1 of the year preceding the meeting.

Science Innovation

In 1992, the AAAS launched an annual conference, alternating between two of the nation's top technology regions: the Pacific Coast and Boston. Sponsored by the AAAS and *Science* magazine, the Science Innovation meeting uniquely focuses on the process rather than the findings of research, showcasing new techniques and instruments. Pioneers of science present plenary overviews of the technologies they have developed; posters and concurrent sessions in the afternoons demonstrate how these technologies can be applied to specific research problems. Extensive industry exhibits and workshops are also featured.

Future Science Innovation Meetings

    1993 Boston (6–10 August)
    1994 Portland (25–29 July)

The Human Genome Conference Series

This annual international conference cosponsored by *Science* magazine and the Human Genome Organization (HUGO) presents plenary sessions and papers by preeminent researchers on the current status and future research directions of human genome research. The conference also includes contributed poster sessions, technical workshops, and numerous scientific exhibits. The Human Genome Conference 1994 will be held in Washington, October 3–4, 1994. For information this meeting, contact Global Trade Productions at (703) 684-1200.

Topical Conferences

In 1992, the AAAS Meetings Office initiated a series of topical interdisciplinary conferences intended for an audience of 500 scientists. The first was "Ion Channels in the Cardiovascular System," cosponsored by AAAS and the National Heart, Lung, and Blood Institute of the National Institutes of Health. Suggestions of topics for future conferences should be submitted to the AAAS Meetings Office at the AAAS address.

Colloquium on Science and Technology Policy

The AAAS Colloquium on Science and Technology Policy, now in its 18th year, is keyed to the federal budget cycle. Organized by the R&D Budget and Policy Program of the Directorate for Science and Policy Programs, the annual conference covers current-year federal research and development budgets (both in the aggregate and by major federal agencies), as well as a wide range of issues in science and technology policy. Recent colloquia have included timely issues such as technology policy and industrial R&D, science and math education, academic R&D, international collaboration, and Soviet and European science and technology developments. Most of the colloquium addresses are published in the annual AAAS Science and Technology Policy Yearbook. **The 1993 meeting was held on 15-16 April.** A one-day colloquium covering similar topics has been held in recent years on the West Coast under the sponsorship of the AAAS Pacific Division. For more information, contact the AAAS Directorate for Science and Policy Programs at the AAAS address.

AAAS Forum for School Science

The AAAS Forum for School Science, a major policy effort in science education, brings together scientists and science educators each year for invited presentations and joint discussions. The meeting's discussions are informed by a companion volume, *This Year in School Science,* which contains review articles and analyses of the topic to be addressed at the conference. In 1991 the conference focused on "Technology: Tools for Teaching and Learning Science and Mathematics." Previous years' topics have included: "Assessment in the Service of Instruction"; "Scientific Literacy"; "Science Teaching: Making the System Work"; and "Students and Science Learning."

Science and Security Colloquium

The AAAS Science and Security Colloquium is held annually in Washington, DC, for scientists, engineers, government officials, security analysts, and educators. The colloquium is organized by the AAAS Program on Science and International Security, Directorate for International Programs, which also prepares an edited volume of original essays related to the conference theme. The 1993 colloquium will be held in October, and will focus on new approaches to security and arms control. For more information, contact the AAAS Program on Science and International Security at the AAAS address.

Regional Division Meetings

The four geographic divisions of the Association hold their own annual meetings; inquiries about these meetings should be directed to the appropriate executive officers (see page 69). Schedules for future meetings are:

Arctic Division
1993—Whitehorse, Yukon Territory, Canada (16-18 September)
1994—Anchorage and the Russian Far East (September)

Caribbean Division
1993—San Juan, Puerto Rico (August/September)
1994—To be announced

Pacific Division
1993—University of Montana, Missoula (20-24 June)
1994—San Francisco State University (19-23 June)

Southwestern and Rocky Mountain Division
1993—University of New Mexico, Albuquerque (23-27 May)
1994—Durango, Colorado (22-25 May)

DIRECTORATES

The special programs of the Association on science education and human resources, international scientific cooperation, and science and policy issues are described below. Each program area is the responsibility of a separate Directorate. For information about any of these programs and activities, write to the Directorate at the AAAS address, or call the number listed on page 2.

Directorate for Education and Human Resources Programs

The Directorate for Education and Human Resources Programs (EHR) seeks to improve the quality of science, mathematics, and technology education (SMT) for all students at all levels; to increase the participation of minorities, women, and people with disabilities in science and engineering; and to improve the public understanding of science and technology for all people. EHR programs focus on supporting systemic educational reform: developing models, materials, mechanisms, processes, and networks; supporting policies, conducting studies and analyses; and implementing findings as appropriate to accomplish the overarching goals.

The directorate moves to support comprehensive efforts that recognize the connectedness of the learning community both in and out of school. To this end, efforts are coordinated through three distinct but collaborating program units: **Science, Mathematics, and Technology Education Programs; Human Resources Programs;** and **Public Understanding of Science and Technology Programs.**

The AAAS Black Church Initiative

This initiative, comprised of several distinct projects, helps churches add enrichment and hands-on science, mathematics, and computer activities to their non-religious educational programs. **The Black Church Project**, funded by the Ford Foundation, links churches across the country with AAAS, corporations, government agencies, academic institutions, community-based organizations, and other professional groups. The **Black Colleges-Black Churches (BC)[2] Project**, funded by the Rockefeller Brothers Fund, is aimed at developing a model program for setting up teaching internships for Black college students in churches that serve the African-American community. Atlanta, Georgia, is the pilot site for the project. The **AAAS Black Church Health Connection Project**, funded by the National Institutes of Health, Public Health Service, is aimed at developing, reviewing, and field-testing hands-on participatory biology activities that promote a healthful lifestyle for use in church-based programs.

The AAAS Forum for School Science

The **AAAS Forum for School Science,** a major policy effort in science education, brings together scientists and science educators each year for invited presentations and joint discussions. The meeting's discussions are informed by a companion volume, *This Year in School Science*, which contains review articles and analyses of the topic to be addressed at the conference. Recent topics have included "National Standards for Science Education," "Technology: Tools for Teaching and Learning Science and Mathematics," "Assessment in the Service of Instruction;" and "Science Teaching: Making the System Work."

The Bell Atlantic/AAAS Institute

The **Bell Atlantic/AAAS Institute,** initiated in 1989 for science and technology teachers in the Bell Atlantic service area, is a model program of continuing professional development and materials development in communications and information technology for middle and junior high school teachers. It provides a comprehensive, year-long program that includes for-credit graduate study, mentorship, and in-service training and computer networking. A partnership between AAAS member scientists and engineers in the teachers' communities helps improve instruction and enhance the teachers' professional development. Recent extensions designed to disseminate the program further include a leadership program leading to teacher-developed and -staffed local/regional institutes and a resource book of science and technology activities developed by teachers as a result of the Institute. More than 140 teachers have participated in this Institute funded by the Bell Atlantic Charitable Foundation.

EHR Media Programs

The EHR media unit produces radio programming and videos and provides technical support for other media-based initiatives in the directorate. Current projects include **Science Update,** the award-winning AAAS radio series on the Mutual Broadcasting Network radio stations, which presents scientific subjects in an entertaining, easy-to-understand style. The 90-second stories cover a variety of topics, from dinosaur calls to solar flares, IQ testing to dancing robot bees. The program airs Mondays, Wednesdays, and Fridays on participating Mutual Broadcasting System stations nationwide. Recent Arbitron surveys put their listening audience at well over one million for each 15-minute period. Other media programs include **Kinetic City Super Crew**, a weekly, half-hour science radio show for children (now under development); **Why Is It?**, a three-and-a-half minute show, also on Mutual, that answers science questions phoned in by listeners on a toll-free line; and **Science America**, a series of occasional science features produced for the Voice of America, and broadcast in Europe.

The Earth Explorer Project

Earth Explorer is a comprehensive resource on environmental issues for students in grades 5 through 9. This *Encyclopedia of the Environment* is being developed in an intriguing, interactive multimedia "Compact Disk-Read Only Memory" (CD-ROM) format and will also be published in a parallel multi-volume print version. The project is designed to respond to the priorities set out in the 1990 National Environmental Education Act and to be consistent with the philosophy and goals of science reforms like Project 2061. **Earth Explorer** promotes interdisciplinary science education by weaving the environmental science together with history, economics, geography, social studies, and the humanities. In the electronic version, multiple search paths will lead students through hundreds of interrelated articles, graphics, data manipulations, and role-playing games. AAAS has partnered with Sonic Images Productions in the design and implementation of the **Earth Explorer**, which has been funded by the U.S. Department of Energy, the National Science Foundation, and Ronald McDonald Children's Charity.

The Hispanic Outreach Initiative

This initiative is designed to strengthen science, mathematics, and technology education for Hispanic youth, particularly at the elementary and high school levels. Key activities have included the production of materials for use in schools and communities, including **Proyecto Futuro: Science and Mathematics Activities in English and Spanish; Stepping into the Future: Hispanics in Science and Engineering** (a role model booklet): Making Science and Mathematics Work for Hispanics (a videotape); and **Science and Mathematics Take-Home Kits for Families.** In 1992, a summit of community- and science-based organizations was conveyed, with support from the National Science Foundation, to develop a national action agenda to increase the participation of Hispanics in science and engineering. An executive summary of the summit's findings will be published and disseminated this year.

Mass Media Science and Engineering Fellowship Program

The **Mass Media Science and Engineering Fellowship Program** places talented science and engineering graduate students with newspapers, television, radio stations, and news magazines for the summer. The fellowship program is designed to strengthen the relationship between science and technology and the media, and to enhance the coverage of science and technology issues in the media to improve public understanding of science.

Math Power

Math Power, supported by the Ford Foundation, provides prototypes of school, community, and home-based activities on conceptually challenging mathematics topics. Teams of parents, teachers, and community educators developed sample mathematics activities to improve the competencies and problem-solving abilities of upper elementary and junior high school students. Planning for field testing, evaluating, and disseminating the materials and the Math Power approach nationwide is underway.

Project MOSAIC

The goal of this National Science Foundation-funded project is to provide training and tools for museums to respond to the reality of the changing demographics. Components include materials development and workshops to help science museums develop and implement strategic plans related to equity; diversifying board membership, staffs, and volunteers; providing training for all members of the museum staff; and developing outreach to minority communities. MOSAIC is a joint project with the Association of Science-Technology Centers, Discovery Place (Charlotte, NC), California Museum of Science and Industry, and the National Air and Space Museum.

Public Science Day

Public Science Day occurs each year in conjunction with the AAAS Annual Meeting. **Public Science Day** brings together science resources that are available locally and is designed to help students in grades K-10 understand the increasing importance that science and technology play in their lives. Students involved learn about career and course selections. They take part in workshops and activities conducted by scientists, such as hands-on science and mathematics experiments. **Public Science Day** teaches students how to use local science resources such as aquariums, botanical gardens, science museums, zoos, and science/technology facilities at universities. Science Encounters for Youth is geared toward high school students and acts as a forum to students to meet with scientists, mathematicians, and engineers. There is a special effort directed toward reaching girls, minorities, and students with disabilities at these programs. AAAS Public Science Day has formed the basis for ongoing **Public Science Day** sponsored by local organizations nationwide.

Science Education News

Published eight times per year and mailed to over 7,000 readers, this newsletter informs the science education community about activities relating to school science, mathematics, and technology education carried out by AAAS, its affiliates, and other organizations. It also provides information on resources and opportunities that may be of interest to readers, and serves as an "idea bank" for organizations expanding their school science, mathematics, and technology education activities.

Science Linkages in the Community

This four-year project, funded by the DeWitt Wallace-Reader's Digest Fund, will help EHR expand its work with community-based organizations. Key features include (a) the development of a coordinated community-based science initiative in three sites involving broad partnerships to support school-based learning and (b) the development of the **AAAS Community Science Linkages Institute** to provide training opportunities and technical assistance for community-based science programs across the United States. The Institute will help community-based organizations plan and implement activities with and for parents, scientists, museums, libraries, policymakers, funders, the media, and the general public.

Science Linkages in the Community extends AAAS's efforts, to join community-based, advocacy, and service organizations with scientists and engineers (individually and through their associations) to improve the mathematics and science education of young people. These efforts are designed to target females, those with physical disabilities, and those who are members of racial or ethnic groups that currently are underrepresented in mathematics, science, and engineering. **Science Linkages in the Community** will draw on the experiences of the Linkages Project. Since 1985, the Linkages Project has grown to include coalitions, collaborations, and networks that involve community-based organizations, local community groups, churches, science-based organizations, school districts, and universities. At the present time, the efforts of the Linkages Project bring these groups together to reach young people in 32 states.

Science + Literacy For Health

With funding from the National Institutes of Health, Public Health Service, AAAS has launched a new project that combines traditional adult literacy efforts with science literacy and health information. Because a great deal of science and health information is conveyed through print media, this project intends to improve the scientific literacy of low-reading level adults by identifying and creating culturally sensitive instructional materials for use in literacy programs and community-based substance abuse and mental health education programs. **Science + Literacy for Health** will produce a reading book and workbook for use in literacy instruction, a software database of reviewed materials, a programming guide for libraries and literacy programs, and outreach materials for a national dissemination campaign.

Science, Technology and Disability Initiative

The **AAAS Science, Technology and Disability Initiative** was founded in 1975 to improve the entry and advancement of people with disabilities in science, mathematics, and engineering. A primary function of the project is to link people with disabilities, their families, professors, teachers, and counselors with scientists, mathematicians, and engineers with disabilities who can share their coping strategies in education and career advancement. Under a five-year grant from the National Science Foundation, the project is working to facilitate the pursuit of engineering degrees by students with disabilities who are enrolled in engineering schools across the country. This program is also creating model programs in engineering schools. In tandem with that effort, the project is developing a video showing engineers with disabilities studying, at work, and commuting to and from work or school. The project also gives technical assistance to students with disabilities and advises AAAS-affiliated societies in making their meetings accessible and produces written material facilitating access to science and engineering education and careers.

Senior Scientists and Engineers (SSE)

The **Senior Scientists and Engineers (SSE)** program is a cooperative effort of leading scientific, engineering, and medical societies that was established to make available volunteer scientists and engineers for a community's needs. Volunteers work on projects in areas of science and mathematics education, public policy, and community service. The program is being piloted in the Washington, D.C., area to test the necessary conditions for success, and plans are under development to expand the project nationally, once these have been established.

The Sourcebook For Science, Mathematics, and Technology Education

More than 2,500 programs and organizations involved in science, mathematics, and technology education and the people responsible for these activities are included in this handy reference tool. The **1992 Sourcebook** contains names, addresses, associations, scientific academies, museums, educational research centers, national advisory groups, and state and federal government agencies. The expanded resources sections list programs and activities for students, educators, parents, and community-group leaders, as well as programs that target under-represented groups. Currently, plans are underway to expand the Sourcebook Project and make the information available electronically.

Women's Programs

Girls and Science: In Touch with Technology, a project funded by the Bush Foundation of St. Paul, Minnesota, is developing resource materials and providing training in hands-on science and mathematics for Girl Scout leaders in 14 local councils serving Minnesota and North and South Dakota. Collectively, these councils serve more than 71,000 girls and 18,000 adult leaders. To assist adult leaders in encouraging science interest among the girls and young women with whom they work, the project forms connections between councils and local scientists, engineers, and other science resources. It also encourages local leaders to develop and adapt their own science activities. EHR has also developed a **Girls and Science: Linkages for the Future Initiative,** for use with community-based organizations and groups serving girls and young women. This program provides hands-on science and mathematics materials, training kits, a videotape describing issues related to girls, mathematics, and science; and a quarterly newsletter, **Girls and Science.** Finally, EHR staff have developed and maintained an extensive network of national and international organizations focused on women in science, engineering, and mathematics, and girl-serving youth organizations.

INTERNATIONAL PROGRAMS

The directorate's activities are grouped into five major program areas. In each case, problems requiring intellectual and material resources beyond the means of any single country are addressed through international cooperation. The resulting collaboration with foreign colleagues has led to significant progress in the solution of common problems, and to the strengthening of relationships with the scientific communities of other countries.

Global Change

The **Global Change Project,** works through the worldwide scientific and engineering communities to address a variety of issues such as climate change, ozone depletion, environmental protection, preservation of biological diversity, and the economic and social consequences of global change. During 1991 and 1992 the project hosted seminars to review the spectrum of governmental and non-governmental UNCED-related endeavors then underway. A symposium for the 1993 annual meeting will address "Science, Agriculture, and Environment in the Former Soviet Union." Plans have been made to publish proceedings from this and other sponsored symposia.

Sub-Saharan Africa Program

The **Sub-Saharan Africa Program** activities are currently divided into two focal areas. The "Documentation and Information for Science and Technology" area includes the Project for African Research Libraries which currently provides almost 200 science and humanities journals, primarily published by AAAS-affiliated societies, to research libraries in 38 sub-Saharan African countries. As part of its "Science and Technology for Development" area, the program offers a series of meetings entitled "Forums on Science in Africa." The 1993 forum will focus on "African Women in Science: Leading from Strength." In both instances the overarching goals are capacity building for science, technology, and development in Africa, and the fostering of effective, long-term partnerships between the scientific and academic communities of the U.S. and sub-Saharan Africa. Among the program's publications are NOTES, a bi-annual newsletter of the Research Library project; *Malaria and Development in Africa: A Cross-Sectoral Approach,* the results of its multidisciplinary study for the U.S. Agency for International Development; the proceedings of three "Science in Africa" meetings; and *Electronic Networking in Africa*, the proceedings of a 1992 workshop on science and technology communication networks in Africa.

Western Hemisphere Cooperation

The **Western Hemisphere Cooperation Project (WHC)** promotes cooperative links between scientists and institutions in the countries of the Americas. A major bilateral program provides support for cooperation between U.S. and Chilean scientists. Support is provided for 14 cooperative research projects in a wide variety of disciplines. The new AMIGO (Americas Interhemisphere Geo-Biosphere Organization) Project is concerned with the effects of global change in analogous ecosystems in North and South America. Still other regional cooperation is coordinated through the Interciencia Association (IA), a 14-nation federation of mostly non-governmental scientific associations in the Americas. WHC serves as host for the IA secretariat. IA organizes symposia, sponsors regional programs in biotechnology and biological resources, and publishes the trilingual journal *Interciencia*.

International Scientific Cooperation

The **Science, Engineering, and Diplomacy Fellowships** program places postdoctoral scientists in offices in the U.S. Department of State and U.S. Agency for International Development in an effort to achieve better integration of science and technology into U.S. development assistance and foreign policy. The program is growing rapidly with 37 fellows presently placed in the U.S. and two abroad.

The **Science Attaché and Foreign Science Lecturer Series** staff arranges seminars for foreign science counsellors and attachés, and organizes lectures by distinguished foreign visitors for the Washington, DC science and technology community.

The **Consortium of Affiliates for International Programs (CAIP)** publishes both an annual directory of affiliates and the newsletter *Consortium Notes*, and sponsors symposia on international science activities. In 1994, the second annual AAAS prize to recognize achievement for international scientific cooperation will be awarded (see page 107).

European Projects. Unprecedented changes in Europe present an opportunity for projects within the European Community and with newly democratic countries in East Central Europe and the former Soviet Union (FSU). The recently-launched newsbrief "Scientist to Scientist" disseminates information on scientific society activities to assist FSU science. A project to provide scientific journals to libraries in Russia, Ukraine and Belarus will begin in April 1993.

Program on Science and International Security (PSIS)

In addition to an annual Colloquium on Science and Security (see page 89), the **Program on Science and International Security** staff develop and present congressional seminars and AAAS annual meeting symposia, as well as conduct an active publications program of issue papers, monographs, books, and proceedings. Congressional seminars provide balanced, non-partisan analyses of important international security issues. The symposia for AAAS members have covered such topics as European security, chemical weapons, arms control verification, and issues of proliferation and regional security.

International Workshops on Advanced Weaponry in the Developing World were convened in 1992 and continue in 1993 to address issues related to nuclear and non-nuclear arms proliferation in three regions of concern the Middle East, South Asia, and Northeast Asia. Policy alternatives, taking into account both the technical dimensions of the problem and the limitations and opportunities provided by regional politics, will be identified and their feasibility considered.

The program also administers the **AAAS-Hilliard Roderick Prize** which recognizes significant contributions in the fields of science, arms control, and international security. The fourth annual prize will be awarded at the 1994 AAAS annual meeting (see page 102–103).

PROJECT 2061

Science Literacy for a Changing Future

Project 2061 is a long-term reform initiative to transform K-12 education for the 21st century so that ALL students achieve science literacy. To fulfill the Project's vision of the future, a coordinated set of tools is being developed for a network of reformers.

Its 1989 report, *Science for All Americans* (available from Oxford University Press), outlined what ALL high school graduates should know and be able to do in natural and social science, mathematics, and technology. Working with teams of educators at six sites across the country and university consultants, Project 2061 is designing tools for school districts and other developers to use in creating their own curricula to achieve *SFAA* goals.

These reform tools include *benchmarks for science literacy* at grades 2, 5, 8, and 12; *curriculum models* with alternative approaches to teaching and learning; *blueprints for systemic reform* on related school issues; and a *curriculum-design system* linking Project 2061 tools to extensive resources.

SCIENCE AND POLICY PROGRAMS

The Directorate for Science and Policy Programs furthers AAAS objectives in three program areas where science, government, and society intersect: Science, Technology, and Government; Science and Human Rights; and Scientific Freedom, Responsibility, and Law.

Three committees support the Directorate's work: the Committee on Science, Engineering, and Public Policy (COSEPP); the Committee on Scientific Freedom and Responsibility (CSFR); and the AAAS American Bar Association National Conference of Lawyers and Scientists (NCLS). Members of COSEPP and the AAAS side of NCLS are appointed by the Board. CSFR is unique among AAAS committees, in that it has both Board and Council-appointed members.

Science, Technology, and Government

The Science, Technology, and Government Program is responsible for a number of programs and activities concerning public policy aspects of science and technology.

Central to these activities is the **R&D Budget and Policy Program,** which analyzes R&D funding trends in the federal budget; organizes the annual AAAS Science and Technology Policy Colloquium; and publishes two reports each year covering federal R&D budgets, plus an annual *AAAS Science and Technology Yearbook.* In 1992, the program also published a special report on federal funding for environmental R&D.

The **Committee on Science, Engineering, and Public Policy** reviews the Association's public policy activities, oversees the S&T Policy Colloquium, and sponsors other activities such as symposia and workshops at the AAAS Annual Meeting to help scientists and engineers communicate more effectively with policymakers.

Additional recent activities have included preparation of *Working with Congress: A Practical Guide for Scientists and Engineers,* published by AAAS in 1992, and development of a database of persons active or interested in science and technology policy issues.

The Science, Technology, and Government staff also administer five fellowship programs for scientists and engineers, which are described below.

Science and Human Rights

The Science and Human Rights Program has four major objectives:

First, it seeks to aid foreign scientists, health professionals, and engineers whose internationally recognized human rights have been violated or who have experienced infringements of academic freedom. Since its establishment in 1976, the Program has taken action on behalf of more than 1,000 individuals in over 75 countries.

Second, it applies scientific methods and procedures to the documentation of violations and the protection of human rights. The Program has provided assistance and training in the field of forensic sciences, statistics and information management, genetics, and torture prevention to human rights groups and government organizations around the world.

Third, the Program promotes greater understanding and support for international human rights in the scientific community, particularly among the membership and affiliated societies of the AAAS. In 1993, the Program plans to launch a Human Rights Action Network where AAAS members can participate in letter-writing campaigns on behalf of their scientific colleagues who experience violations of their human rights.

Finally, the Program examines in depth selected issues in the field of human rights to improve the conceptual bases and evaluate the implementation of human rights. Currently, the program is exploring the relationship between environmental protection and human rights, and attempting to define and develop measurements for the right to health care and the right to education.

In May 1992, the Program received the American Psychiatric Association's first annual Human Rights Award, in recognition of its pioneering work in science and human rights.

Scientific Freedom, Responsibility and Law

The Scientific Freedom, Responsibility and Law Program is committed to upholding standards of science and engineering among scientists and engineers and to improving understanding and sensitivity to issues of professional ethics among scientists. In addition, it works to promote a better understanding of science among lawyers and judges and of the legal system among scientists; and it monitors emerging ethical, legal, and social issues related to science and technology and brings them to the attention of policymakers and the scientific community.

In support of these objectives, the Program sponsors projects, conferences, and publications addressing such issues as: (1) the ethical and legal implications of genetic testing; (2) minority perspectives on values and ethics in science and technology; (3) the use of animals in research and education; (4) fraud and misconduct in science; (5) the use of scientific and technical information in the courts; (6) the ethical and legal aspects of computer network use and abuse; and (7) the effects of national security controls on unclassified research.

The **Committee on Scientific Freedom and Responsibility** and the **National Conference of Lawyers and Scientists** work closely with Program staff to promote the goals and programs described above.

The Program also coordinates the **Professional Society Ethics Group,** which provides an opportunity for representatives from professional societies to exchange information and ideas related to professional ethics, and publishes the quarterly newsletter, *Professional Ethics Report.*

Science and Engineering Fellowship Programs

The directorate administers several fellowship programs that provide opportunities for scientists and engineers to work on public policy issues in Congress and federal agencies. These include:

The AAAS Congressional Science and Engineering Fellowship Program, which places Fellows either in Members' offices or with congressional committees for one year.

The AAAS/Sloan Executive Branch Science and Engineering Fellowship Program, which places industry-based Fellows for one or two years in the White House Office of Science and Technology Policy.

The AAAS Science, Engineering, and Diplomacy Fellowship Program which enables Fellows to work for a year at either the U.S. Agency for International Development or the Department of State, including assignments abroad under a new **Overseas Diplomacy Fellowship Program.**

The AAAS/EPA Environmental Science and Engineering Fellowship Program, which places Fellows at the U.S. Environmental Protection Agency for a summer program of research on environmental issues.

AWARDS, PRIZES, AND GRANTS

To recognize scientists, journalists, and public servants for significant contributions to science and to the public's understanding of science, AAAS administers the awards listed below. (See pages 14–17 for current members of the award committees.) Nominations for each prize must be received by the cutoff dates indicated and should be addressed to the Office or Directorate responsible for the selection process. All awards are presented at the Annual Meeting immediately following the award year. In addition, all awardees are reimbursed for travel and one night's hotel expenses incurred in attending the award presentation.

AAAS–Philip Hauge Abelson Prize

The AAAS–Philip Hauge Abelson Prize of $2,500 and a commemorative plaque, established by the AAAS Board of Directors in 1985, is awarded annually either to a public servant in recognition of sustained exceptional contributions to advancing science or to a scientist whose career has been distinguished both for scientific achievement and for other notable services to the scientific community.

AAAS members are invited to submit nominations, each of which must be endorsed by at least two other AAAS members. Nominations should be typed and should include the following information: nominee's name, institutional affiliation and title, address and biographical resume; statement of justification for nomination; and names, identification, and signatures of three or more sponsors. The winner is selected by a seven-member selection panel.

Nominations for the 1993 prize should be submitted to AAAS–Philip Hauge Abelson Prize, Directorate for Science and Policy Programs, AAAS, 1333 H St., N.W., Washington, DC 20005, for receipt on or before 1 August 1993.

Recipients of the Prize

1985	**James A. Van Allen**, University of Iowa, Iowa City.
1986	**James A. Shannon**, National Institutes of Health (retired).
1987	**Norman Hackerman**, Rice University.
1988	**John T. Edsall**, Harvard University (emeritus).
1989	**Franklin A. Long**, University of California at Irvine; formerly at Cornell University.
1990	**George E. Brown, Jr.**, U.S. Congress
1991	**Bentley Glass**, State University of New York at Stony Brook (emeritus).
1992	**John H. Gibbons**, Assistant to the President for Science and Technology, and Director, White House Office of Science and Technology Policy; formerly Director, Office of Technology Assessment, U.S. Congress.

AAAS–Newcomb Cleveland Prize

The Association's oldest award, established in 1923 with funds donated by Newcomb Cleveland of New York City, was originally called the AAAS Thousand Dollar Prize. It is now known as the AAAS–Newcomb Cleveland Prize, and its value has been raised to $5,000. The winner also receives a bronze medal.

The prize is awarded to the author or authors of an outstanding paper published in the Articles, Research Articles, or Reports sections of *Science*. Each annual contest starts with the first issue of June and ends with the last issue of the following May.

An eligible paper is one that includes original research data, theory, or synthesis; is a fundamental contribution to basic knowledge or a technical achievement of far-reaching consequence; and is a first-time publication of the author's own work. Reference to pertinent earlier work by the author may be included to give perspective.

Throughout the year, readers of *Science* are invited to nominate papers appearing in the Articles, Research Articles, or Reports sections. Nominations must be typed and the following information provided: title of the paper, author's name, date of issue in which it was published, page number, and a brief statement of justification for nomination. Final selection is by a panel of distinguished scientists appointed by the editor of *Science*.

Nominations for the current contest should be submitted by 30 June 1993 to AAAS–Newcomb Cleveland Prize, AAAS/*Science*, Rm. 924, 1333 H St., N.W., Washington, DC 20005.

Recipients of the Prize

A complete list of recipients is available from *Science* magazine. Winners since 1986 have been:

1986–1987	**Jeremy Nathans, David S. Hogness, Darcy Thomas, Thomas P. Piantanida, Roger L. Eddy,** and **Thomas B. Shows** for two articles on molecular genetics of human color vision, 11 April 1986, and **Arthur J. Zaug** and **Thomas R. Cech** for a research article, "The Intervening Sequence RNA of *Tetrahymena* Is an Enzyme," 31 January 1986.
1987–1988	**Mario J. Molina, Tai-Ly Tso, Luisa T. Molina,** and **Frank C.-Y. Wang** for a research article, "Antarctic Stratospheric Chemistry of Chlorine Nitrate, Hydrogen Chloride, and Ice: Release of Active Chlorine," and **Margaret A. Tolbert, Michel J. Rossi, Ripudaman Malhotra,** and **David M. Golden** for a report, "Reaction of Chlorine Nitrate with Hydrogen Chloride and Water at Antarctic Stratospheric Temperatures," 27 November 1987.
1988–1989	**William H. Landschulz, Peter F. Johnson,** and **Steven L. McKnight** for two research articles, "The Leucine Zipper: A Hypothetical Structure Common to a New Class of DNA Binding Proteins," 24 June 1988, and "The DNA Binding Domain of the Rat Liver Nuclear Protein C/EBP Is Bipartite," 31 March 1989.
1989–1990	**Margaret J. Geller** and **John P. Huchra** for the article, "Mapping the Universe," 17 November 1989.
1990–1991	**Stephen P.A. Fodor, J. Leighton Read, Michael C. Pirrung, Lubert Stryer, Amy Tsai Lu,** and **Dennis Solas** for the research article "Light-Directed, Spatially Addressable Parallel Chemical Synthesis," 15 February 1991.
1991–1992	**Paul D. Quay, Bronte Tilbrook** and **C.S. Wong** for the research article, "Oceanic Uptake of Fossil Fuel CO_2: Carbon-13 Evidence," 3 April 1992.

AAAS Award for Behavioral Science Research

Established in 1952 with funds donated by Arthur F. Bentley, the AAAS Award for Behavioral Science Research (formerly the Socio-Psychological Prize) of $2,500 is offered annually for a meritorious paper in the behavioral sciences. It is awarded for innovative studies and analyses that (1) further understanding of human psychological-social-cultural behavior and (2) foster liberation from philosophic-academic conventions and from dogmatic boundaries between different disciplines. Entries should deal with basic observation and construction in social process, group behavior, or interpersonal behavior.

The purpose of the award is to encourage the development and application of methods for the study of social behavior, using the logic of observation and explication so fruitful in any scientific endeavor. The research reports submitted should be based on explicitly stated assumptions or postulates that lead to conclusions or deductions that are verified by systematic empirical research. Papers should present a completed analysis of a problem, the relevant data, and an interpretation of the data in terms of the assumptions with which the study began. Entries must have been published in peer-reviewed journals since 1 January 1992.

The deadline for receipt of entries is 1 July 1993. For entry blank and instructions, write to Catherine Campos, AAAS Award for Behavioral Science Research, Directorate for Education and Human Resources Programs, AAAS, 1333 H St., N.W., Washington, DC 20005.

Recipients of the Award

A list of earlier recipients is available from the Directorate. Winners since 1987 have been:

1987 **Philip E. Tetlock**, "Monitoring the Integrative Complexity of American and Soviet Foreign Policy Rhetoric: What Can Be Learned?"
1988 **Leda Cosmides**, "The Logic of Social Exchange: Has Natural Selection Shaped How Humans Reason?"
1989 **Cathy Spatz Widom**, "The Cycle of Violence."
1990 **Bruce P. Dohrenwend, Itzhak Levav, Patrick E. Shrout, Sharon B. Schwartz, Guedalia Naveh, Bruce G. Link, Andrew E. Skodol,** and **Ann Stueve**, "Socioeconomics Status and Psychiatric Disorders: A Test of the Social Causation-Social Selection Issue."
1991 **Gerd Gigerenzer**, "From Tools to Theories: A Heuristic of Discovery in Cognitive Psychology."
1992 **Patricia Marks Greenfield**, "Language, Tools, and Brain: The Ontogeny and Phylogeny of Hierarchically Organized Sequential Behavior."

AAAS–Hilliard Roderick Prize

The AAAS–Hilliard Roderick Prize for Excellence in Science, Arms Control, and International Security acknowledges recent outstanding contributions to the understanding of critical issues related to arms control and international security having an important scientific or technical dimension. Technology development, scholarly publications, timely analyses, or other professional activities that have advanced thinking about arms control and international security are recognized.

The prize of $5,000 and a commemorative medal will be presented at the 1994 Annual Meeting. Nominations should include a letter describing the recent contribution and its importance, two supporting letters, and additional supporting documentation. Nominations should be submitted to AAAS–Hilliard Roderick Prize, Directorate for International Programs, AAAS, 1333 H St., N.W., Washington, DC 20005, for receipt on or before 13 September 1993.

Recipient of the Award

1993 **Dr. Sidney Drell**, Professor and Deputy Director, Stanford Linear Accelerator Center, Stanford University was cited for his exemplary efforts in applying technical expertise to complex national security issues, such as those related to the Strategic Defense Initiative, possible elimination of ballistic missiles, the appropriate limits on nuclear testing, and the verification of major arms control.

AAAS Scientific Freedom and Responsibility Award

The AAAS Scientific Freedom and Responsibility Award, which consists of $2,500 and a plaque, was established in 1981 to honor scientists and engineers whose exemplary actions, often taken at significant personal cost, have served to foster scientific freedom and responsibility.

Eligible nominees are scientists and engineers who have acted to protect the public's health, safety, or welfare; or to focus public attention on potentially serious impacts of science and technology on society by their responsible participation in public policy debates; or to establish important new precedents in social responsibility or in defending the professional freedom of scientists and engineers.

Nominations for the 1994 award must be received by 15 July 1993. For nomination forms and instructions, write to AAAS Scientific Freedom and Responsibility Award, Directorate for Science and Policy Programs, AAAS, 1333 H Street, N.W., Washington, DC 20005.

Recipients of the Award

A list of earlier recipients is available from the Directorate. Winners since 1987 have been:

1987 **Francisco J. Ayala**, for speaking out forcefully on the scientific basis for the theory of evolution, while seeking a foundation of scientific knowledge and logic that demonstrates a common ground for evolutionary theory and ethical values.

Norman D. Newell, for his early and prescient campaign to alert scientists to the importance of public understanding of the theory of evolution and to the threats creationism poses to academic freedom and science education.
Stanley L. Weinberg, for his leadership in mobilizing local opposition among scientists and teachers to challenges the creationist movement poses to the integrity of science and the teaching of evolution.

1988 **Roger M. Boisjoly**, for his exemplary and repeated efforts to fulfill his professional responsibilities as an engineer by alerting others to life-threatening design problems of the *Challenger* space shuttle and for steadfastly recommending against the tragic launch of January 1986.

Richard L. Garwin, for his courageous, sustained, and effective efforts over a distinguished scientific career to educate government policy makers and the public on such highly controversial applications of science to technology as the Anti-Ballistic Missile System, the Supersonic Transport, and the Strategic Defense Initiative.

1989 **Natural Resources Defense Council**, for a bold and successful initiation of a nuclear test ban verification project requiring in-country monitoring in both the United States and the Soviet Union.

Robert L. Sprague, for his courage and persistence in reaffirming the highest standards of scientific integrity by initiating the censure of a research colleague who fabricated data while conducting experiments with mentally retarded patients.

1990 **Matthew S. Meselson**, for chairing the AAAS Herbicide Assessment Commission, influencing the Nixon administration to renounce biological warfare and to unilaterally destroy the U.S. stockpile of biological weapons, and challenging claims by the U.S. government that the Soviets were supporting the use of a yellow, rain-like chemical warfare agent in Southeast Asia.

1991 **Adrian R. Morrison**, for his dedicated promotion of the responsible use of animals in research and his courageous stand in the face of great personal risk against attempts to curtail animal research essential to public health.

1992 **Inez Austin**, for her courageous and persistent efforts to prevent potential safety hazards involving nuclear waste contamination at the Hanford Nuclear Reservation.

1993 **Daniel L. Albritton**, for leadership in organizing research on stratospheric ozone destruction and in responsibly conveying the knowledge to the international negotiations designed to protect the Earth's vital ozone shield.

Robert T. Watson, for leadership in developing the foundation for international protocols for control of ozone-destroying chemicals and for protection of the Earth's vital ozone shield.

AAAS Award for Public Understanding of Science and Technology

The AAAS Award for Public Understanding of Science and Technology, established in 1987, recognizes working scientists and engineers who make outstanding contributions to the "popularization of science." The award, sponsored by AAAS carries a $5,000 prize and a commemorative plaque.

Nominations for the 1993 award should be sent by 1 August 1993 to AAAS Award for Public Understanding of Science and Technology, c/o Judy Kass, Committee on Public Understanding of Science and Technology, AAAS, 1333 H St., N.W., Washington, DC 20005.

Recipients of the Award

1987 **Philip Morrison,** professor emeritus of physics at MIT, for his exemplary, life-long commitment to teaching science and reaching wide audiences through the popular media.

1988 **Anthony M. Fauci, M.D.,** for exceptional contributions to public under standing of immunology and, specifically, AIDS, which have helped calm hysteria, put new findings in perspective, and increase public confidence in the scientific process.

1989 **Robert D. Ballard,** for his innovative contributions to public understanding of the partnership between science, technology, and the social sciences, especially as they relate to underwater discoveries.

1990 **William L. Rathje,** University of Arizona, for his innovative contributions to public understanding of science and its societal impacts by demonstrating with his creative "Garbage Project" how the scientific method can document problems and identify solutions.

1991 **Stephen H. Schneider,** National Center for Atmospheric Research, for furthering public understanding of environmental science and its implications for public policy through writing, lecturing, and working with professional societies, government institutions, and the media.

1992 **Farouk El-Baz,** Boston University, for furthering public understanding of science related to arid lands through his innovative use of space photography and easy-to-understand language.

AAAS–Westinghouse Science Journalism Awards

With support from the Westinghouse Foundation, the Association offers five annual AAAS–Westinghouse Science Journalism Awards of $2,500 each to encourage and recognize outstanding reporting on the sciences and their applications, excluding health and clinical medicine. The awards are for reporting in (1) newspapers of over 100,000 daily circulation, (2) newspapers of under 100,000 circulation, (3) general circulation magazines, and on (4) television and (5) radio.

The contest year for 1993 is 1 July 1992–30 June 1993. Entries are due at the AAAS by 13 July 1993. For entry rules and instructions, write to AAAS–Westinghouse Science Journalism Awards, Communications Office, AAAS, 1333 H St., N.W., Washington, DC 20005.

A complete list of previous recipients is available from the AAAS Office of Communications. Winners since 1988 are listed below.

Recipients of the Award–Newspapers (over 100,000 daily circulation)

1988	**Gayle Golden,** *Dallas Morning News.*
1989	**Tom Siegfried,** *Dallas Morning News.*
1990	**Charles Petit,** *San Francisco Chronicle.*
1991	**Robert Cooke, B.D. Colen, Earl Lane, Peter Marks,** and **Laura Muha,** *Newsday*
1992	**Natalie Angier,** *The New York Times* (tie) **Deborah Blum,** *The Sacramento Bee*

Recipients of the Award–Newspapers (under 100,000 circulation)

1988	**Tom Foster** and **Matthew Cox,** *Syracuse Post-Standard.*
1989	**Jim Borg,** *Honolulu Advertiser.*
1990	**Lawrence Spohn,** *Albuquerque Tribune.*
1991	No award given.
1992	**Jim Kelly,** *Houston Press*

Recipients of the Award–Magazines

1988	**John Tierney, Karen Springen,** and **Lynda Wright,** *Newsweek.*
1989	**Hans Christian von Baeyer,** *The Sciences.*
1990	**Michael Lemonick, Philip Elmer-DeWitt, J. Madeleine Nash,** and **Christopher Redman,** *Time.*
1991	**John Horgan,** *Scientific American.*
1992	**Richard Preston,** *The New Yorker*

Recipients of the Award–Radio

1988	**Richard Harris** and **Michael Skoler,** National Public Radio.
1989	No award given.
1990	No award given.
1991	**David H. Baron,** WBUR-FM.
1992	**Larry Massett,** SOUNDPRINT

Recipients of the Award–Television

1988	**Robert Hone,** KQED-TV (San Francisco).
1989	**Lionel Friedberg, Georgann Kane,** and **Linda K. Reavely,** WQED-TV (Pittsburgh).
1990	**Larry Engel, Thomas Lucas,** and **Paula Apsell,** NOVA.
1991	**William Kurtis,** Kurtis Productions, Ltd. **Leslie Reinherz,** Chedd-Angier Productions Co.
1992	**Paula Apsell and Tom Levenson,** WGBH/NOVA

AAAS Mentor Award

The Board of Directors of the American Association for the Advancement of Science has established an award, to be given annually, to honor individuals who, during their careers demonstrated extraordinary leadership in efforts above and beyond their duties to increase the participation of women, minorities, and people with disabilities in science and engineering.

The award, consisting of a plaque and $5,000, is awarded to individuals who have affected the climate of a department, college, or institution in such a way as to significantly increase the ethnic diversity of students pursuing and completing doctoral studies, and who have mentored and guided to the completion of doctoral studies significant numbers of minority, female, and disabled students.

Nominees would have assisted students in many, if not all, of the following: encouraged and helped them find an appropriate graduate school; accompanied the students on their initial meetings with their major professors; introduced the students to the graduate environment, and socially to the professionals; helped them find financial support; provided psychological support and encouragement; helped the students to present and publish their work; provided career guidance and assistance in placement in a meaningful post-doctoral position and employment; continued interest in the individual's professional advancement.

Nominators should send a narrative statement about the mentor, the nominee's curriculum vita, and sufficient documentation to show evidence that the nominee meets the criteria, such as support letters from students and colleagues.

Materials should be sent to: AAAS Mentor Award, c/o Yolanda S. George, Directorate for Education and Human Resources Programs, AAAS, 1333 H Street N.W., Washington, DC 20005, Fax 202-371-9849.

Applications will be reviewed and award recommendation made by the AAAS Committee on Opportunities in Science. The awardee will be notified by December 15, 1993, and will be expected to participate in the February award ceremony.

Recipients of the Award

1992   **Anthony J. Andreoli,** California State University, Los Angeles, CA.
          Lafayette Frederick, Howard University, Washington, DC.
1993   **Abdulalim Abdullah Shabazz,** Clark Atlanta University, Atlanta, GA.

AAAS International Scientific Cooperation Award

Many scientists and engineers contribute valuable time away from the established career paths of research, teaching, and publishing to foster activities and develop programs of an international nature. AAAS, in collaboration with its affiliated organizations, seeks to recognize an individual or a limited number of individuals working together in the scientific or engineering community for making an outstanding contribution to furthering international cooperation in science and engineering.

A monetary prize of $2,500, a certificate of citation, and travel expenses to attend the AAAS annual meeting to receive the award are included.

Nominations for the award may be originated by anyone submitting a letter of recommendation on or before September 15 and directed to the AAAS International Scientific Cooperation Award, Directorate for International Programs, AAAS, 1333 H St., N.W., Washington, DC 20005.

AAAS Academy Research Grants

The Association annually makes available to the affiliated academies of science funds for use in promoting student research. Research funds are awarded on a selective basis to academies for projects that encourage secondary school students to do research. For further information, write to Directorate for Education and Human Resources Programs, AAAS, 1333 H St., N.W., Washington, DC 20005.

AAAS William D. Carey Annual Science Award

The AAAS William D. Carey Annual Science Award was established in 1989 to encourage advanced students in science or engineering to participate in the AAAS Annual Meeting. This award provides up to $300 support for students to attend the meeting. In addition, AAAS will furnish registration fees and a year's subscription to *Science*.

Graduate students are invited to apply for the 1994 William D. Carey Annual Science Award. Consideration for this award will be based on academic achievement and other accomplishments related to the advancement of science. Applicants should submit a curriculum vitae and a 200–300 word description of their research topics to Amie Hubbard; AAAS; 1333 H Street, N.W.; Washington, D.C., 20005. Deadline for receipt of applications is November 1, 1993.

Recipients of the Award:

1990	**David D. Norris**, Tulane University	
	Meredith Mason Garcia, Tulane University	
	S. Levi Taylor, Tulane University	
1991	**Marijke L. Bekken**, University of California	
	Fernando E. Vega, University of Maryland	
	Robert S. Williamson, Jet Propulsion Laboratory	
	Alan Dardik, University of Pennsylvania, School of Medicine	
	Kiumars Lalezarzadeh, SUNY–Stony Brook	
1992	**Patrick M. Doogan**, DePaul University	
	Julie Eaker, DePaul University	
1993	**Wesley Hawkins**, Howard University	
	Teresa Stevenson, Howard University	

III

History
Policy
Governance

History

FOUNDING

The founding of the American Association for the Advancement of Science in Philadelphia in 1848 was the first successful effort to establish in the United States a national scientific society embracing all the sciences, but not the first effort to bring such a society into being.

In 1816, John Quincy Adams, John C. Calhoun, Daniel Webster, Edward Everett, Henry Clay, and other national leaders launched the Columbian Institute for the Promotion of Arts and Sciences. Its objectives included the advancement of science, the establishment of several national institutions, and the promotion of agriculture and foreign commerce. The Columbian Institute was started with high hopes, but the founders were preoccupied with other activities. By 1825 it became inactive, and in 1840 it "passed into" the National Institution for the Promotion of Science.

The National Institution was incorporated by act of Congress in 1840. Monthly meetings were held for a year or two; three bulletins were published; and in April 1844, the National Institution sponsored a Congress of Scientists in Washington, to which it invited members of other learned societies. After this congress, the National Institution never held another meeting, although technically it continued to exist until 1861.

In 1831, the British Association for the Advancement of Science was founded. In 1837, John Collins Warren, a prominent medical scientist of Boston, read a paper at one of its sessions and was impressed with the value of one large meeting devoted to all the sciences. Upon his return, he endeavored to found a parallel association in America. He invoked the cooperation of the American Philosophical Society, but that organization decided his idea was "inexpedient."

Despite these failures, the first half of the nineteenth century was a time of scientific advance: Dalton's atomic theory (1808); Avogadro's hypothesis (1811); the synthesis of urea by Wöhler (1828); Lyell's *Principles of Geology* (183033); the cell theory of Schleiden and Schwann (183839); the work of Faraday and Henry on electricity and magnetism (from 1820 on) are examples.

Geology was particularly active. The mineral resources of a vast continent were largely unexplored. Between 1823 and 1839, 17 states began geological surveys. In 1819, geologists incorporated the American Geological Society. One of its officers was Benjamin Silliman of Yale, who in 1818 had started the long-lived and important American Journal of Science and Arts. Although the American Geological Society lapsed in 1826, its spirit reappeared a few years later in the Association of American Geologists, which, again a few years later and after a change of name, was transformed into the American Association for the Advancement of Science.

In 1837, and probably earlier, Edward Hitchcock of Amherst College, who had been a member of the American Geological Society, advocated a national association of geologists. Several geologists, among them Lardner Vanuxem and James Hall, had discussed such an association. Crediting Hitchcock with the idea, W. W. Mather arranged a meeting on 20 November 1839, at the Albany home of Ebenezer Emmons, who was doing a portion of New York's geological survey. At that meeting, Vanuxem, Hall, Mather, Timothy A. Conrad, and Emmons drew up plans for a national association, and Vanuxem conducted the subsequent correspondence. A second conference of the same group was held in 1839 at the Emmons house, Hudson Avenue and High Street (in 1901, AAAS placed a commemorative tablet at this site). As a result, the Association of American Geologists was organized on 2 April 1840, at the Franklin Institute, Philadelphia, by 18 prominent geologists of seven eastern states. Edward Hitchcock was chairman.

In 1842, at Boston, this society became "The Association of American Geologists and Naturalists." The new name was logical, for many of the geological surveys of the period included zoology and botany as well. As the number of scientists increased and scientific disciplines became more specialized, the need for a national organization that would include all the sciences became more apparent. When the Association of American Geologists and Naturalists met in Boston in September 1847, under the chairmanship of William Barton Rogers, the society resolved to reorganize and to change its name to the "American Association for the Promotion of Science... designed to embrace all labourers in Physical Science and Natural History." Officers for the new American Association were elected, and a committee of three (H. D. Rogers, Benjamin Peirce, and Louis Agassiz) was appointed to draw up a new "Constitution and Rules of Order" and to report at the meeting set for Philadelphia in 1848.

On 20 September 1848, the American Association for the Promotion of Science met in the Academy of Natural Sciences in Philadelphia. William B. Rogers of the University of Virginia, the last president of the Association of American Geologists and Naturalists, read the report of the committee on the constitution. The committee proposed a slightly different name for the new organization and recommended that:

> The Society shall be called THE AMERICAN ASSOCIATION FOR THE ADVANCEMENT OF SCIENCE. The objects of the Association are, by periodical and migratory meetings, to promote intercourse between those who are cultivating science in different parts of the United States; to give a stronger and more general impulse, and a more systematic direction to scientific research in our country; and to procure for the labours of scientific men, increased facilities and a wider usefulness.

The organizational meeting then adjourned to reconvene in the "Hall of the University of Pennsylvania," at 4 p.m. on the same day, as the first scientific session of the AAAS. William Rogers introduced the Association's first elected president, William C. Redfield of New York. Among the eminent scientists present were Louis Agassiz, Stephen Alexander, Alexander D. Bache, Asa Gray, Joseph Henry, Benjamin Silliman, and John Torrey—names still revered today. At this first meeting of the Association, which lasted five days, Benjamin Peirce read a paper "On the General Principles of Analytical Mechanics." He was followed by Louis Agassiz, whose paper was "On the Classification of the Animal Kingdom." Altogether some 60 papers were presented. Apparently, 87 of the 461 charter members were present.

In 1848, the Association had two sections, "Natural History, Geology, etc." and "General Physics, etc."; by 1881, there were nine sections; now there are 22. The charter membership of 461 has expanded to 132,000. The objectives, however, have remained fundamentally as stated by the founders 139 years ago.

A definitive history of the American Association for the Advancement of Science has yet to be written. The historical portion of the *Summarized Proceedings and Directory, 1940–48*, written by Forest Ray Moulton, is a useful summary; Dael Wolfle's *Renewing a Scientific Society: The American Association for the Advancement of Science from World War II to 1970* is a detailed discussion of all aspects of AAAS activities; and the founding years are covered exhaustively by Sally Gregory Kohlstedt, *The Formation of the American Scentific Community: The American Association for the Advancement of Science, 1848-1860* (University of Illinois Press, 1976). Some of the events of past years are summarized here.

MILESTONES OF THE AAAS

1840 The Association of American Geologists, lineal ancestor of the AAAS, was organized on 2 April at the Franklin Institute, Philadelphia. Amherst professor of chemistry and natural history, Edward Hitchcock, who had advocated a general association of scientists, was the first chairman.

1842 It was decided to change the name of the Association of American Geologists to the Association of American Geologists and Naturalists, in recognition of the breadth of interests of the membership.

1847 At its meeting in Boston on 24 September, the AAGN passed a resolution to reorganize as the American Association for the Promotion of Science; a committee was appointed to draft a constitution.

1848 On 20 September, at the Academy of Natural Sciences of Philadelphia, the AAGN formally became the AAAS, with 461 charter members. The first meeting lasted five days; some 60 papers were read on Natural History and General Physics before the two sections, chaired, respectively, by Louis Agassiz and Joseph Henry.

1850 The office of "Permanent Secretary" was established. Spencer F. Baird was elected to begin service the following year at an annual salary of $300.

1851 The "Objects and Rules of the Association" were accepted and thereafter known as the first AAAS constitution. Dues were increased from $1 to $2; purchase of the *Proceedings* for an additional dollar was optional.

1852 No meeting of the young AAAS was held because of "the prevalence of cholera along the approaches to Cleveland from the south."

1856 A new (second) constitution was adopted. Since there were no bylaws, the *Proceedings,* for a period, cumulatively listed the resolutions of the legislative body, which was called the Standing Committee.

1857 The office of vice president for each meeting was established. (President J. W. Bailey, at 46, had died in office before the First Montreal Meeting.)

1861-1865 Because of the Civil War, there were no meetings for five years, and no presidents were elected.

1874 In March, the Association was incorporated under the laws of Massachusetts. A new (third) constitution of 39 articles was adopted; it provided that each section (then two) should have a vice president, one of whose duties would be to give an annual address. Provision was made for the nomination of Fellows from among members who were professional scientists or significantly advancing science. Membership dues were increased from $2 to $3, with the *Proceedings* included; and life memberships were offered at $50. The "admission fee" of $5, begun in 1867, was continued.

1881   An amendment to the constitution recognized nine sections, each with a vice president.

1887   The Standing Committee was renamed the Council.

1893   The section on Biology, which had absorbed the Section on Microscopy in 1886, was divided into Zoological Sciences (F) and Botanical Sciences (G).

1895   The Council appointed a committee to study AAAS policies. The Policy Committee continued as a standing committee for some years, gradually acquiring specific duties, such as for meetings and publications (1904).

1899   A constitutional amendment provided that members of affiliated societies could attend AAAS meetings if they registered. This was the first mention of "affiliated societies" in the constitution, but the Council had used the term in 1895, and a few societies had met with AAAS as early as 1891.

1900   *Science,* owned and edited by James McKeen Cattell, was made the official publication of the Association, and thereafter was sent to all members.

1901   A constitutional amendment provided that affiliated societies would be represented on the Council by one or two persons, according to the number of AAAS Fellows in their memberships.

1902   Because of academic summer schools and to accommodate more affiliated societies, the annual meeting date changed from August to late December.

1907   The *Proceedings* ceased to carry the details of papers and addresses, since many of them were being published in *Science*. The headquarters of the Association, which had been in the office or home of the permanent secretary, was located in the Smithsonian Institution.

1908   This was the last year the *Proceedings* were published as an annual volume. *Summarized Proceedings,* covering more than one year's meetings, appeared from 1910 to 1948.

1915   The Pacific Division of the AAAS was established; its first meeting was held in 1916 in San Diego. *The Scientific Monthly* became an official journal of the Association.

1919   A new (fourth) constitution of 11 articles and 11 bylaws provided for 15 sections of the Association.

1920   The Southwestern and Rocky Mountain Division of the AAAS was established and held its first meeting in El Paso. The Policy Committee was renamed the Executive Committee.

1923   The first AAAS Thousand Dollar Prize (now called the Newcomb Cleveland Prize) was awarded to Leonard Eugene Dickson for his paper on "The Theory of Numbers."

1924   The Annual Exposition of Science (now called the AAAS Exhibition) was established as an organized and integral feature of the annual meetings of the Association. There had been a few commercial exhibits earlier.

1934   The first of a series of special monographs, The Protection by Patents of "Scientific Discoveries," was published as a supplement to *Science*. Since

1938, books in the series have been designated as symposium volumes.

1938 The Gibson Island Research Conferences, started in 1931 by Neil Gordon, were established on a permanent basis under AAAS auspices. In 1948 they were named the Gordon Research Conferences to honor their founder.

1942-1943 No meetings were held because of World War II. From March 1942 to December 1946, the *Bulletin* was issued monthly to keep members informed of Association activities. The Association purchased *The Scientific Monthly* in 1943 from James McKeen Cattell.

1944-1945 The Association purchased *Science* in 1944 from the estate of James McKeen Cattell. The 1944 meeting was held in September in Cleveland under wartime conditions. Because of travel restrictions, the 1945 meeting was delayed until March 1946.

1946 A new (fifth) constitution with no bylaws redefined the objects of the Association, defined the powers of the Council, and provided for a president-elect. On 9 September, the AAAS moved from the Smithsonian Institution into a large old house near Scott Circle, Washington, DC.

1947 The Section on Medical Sciences (N) was divided into three subsections: Medicine, Dentistry, and Pharmacy.

1948 During 13-17 September in Washington, DC, the Association celebrated its centenary with a meeting on the theme of "One World of Science."

1951 The Alaska Division of the AAAS was established and held its first meeting at Mount McKinley National Park. The Executive Committee issued the "Arden House Statement," reassessing programs and policies. The Section on Industrial Science (P) was authorized.

1952 A new (sixth) constitution of 12 articles and 11 bylaws changed the Executive Committee to a Board of Directors.

1954 The Subsection on Medicine became the Section on Medical Sciences, and the Subsections on Dentistry (Nd) and Pharmacy (Np) became sections in their own right. (The Section on Pharmacy was renamed the Section on Pharmaceutical Sciences in 1962.)

1955 Ground was broken in April for a headquarters building on Scott Circle; temporary quarters were rented at 1025 Connecticut Avenue. The Association sponsored an International Conference on Arid Lands in New Mexico, 26 April-4 May. The AAAS Socio-Psychological Prize, endowed by Arthur F. Bentley, was established on an annual basis. (In 1985, the award was renamed the AAAS Prize for Behavioral Science Research.) The AAAS Science Teaching Improvement Program was initiated.

1956 The new three-story headquarters building at 1515 Massachusetts Avenue was occupied on 25 May.

1958 *The Scientific Monthly* was merged into *Science*. A "Parliament of Science" on national science policy was held in Washington, DC.

1959 The first International Oceanographic Congress was held under AAAS auspices in New York City. The Symposium on Basic Research was also held in New York City. The AAAS Westinghouse Science Writing Awards were reestablished. (An earlier series had been administered from 1946 through 1953.) The

116 / HISTORY

	Committee on Public Understanding of Science was begun.
1960	Amendments to the constitution and bylaws defined more clearly the responsibilities and duties of the Council and Board of Directors. A Committee on Council Affairs was established.
1961	Publication of the *Bulletin* was resumed on a quarterly basis. Establishment of two new sections of the Association was authorized: Information and Communication (T) and Statistics (U).
1962	With support from the National Science Foundation, a Commission on Science Education was established to direct programs and develop materials for science instruction and the education of science teachers.
1965	*Science Books: A Quarterly Review* was established "to improve ...science education... and the public understanding of science" by critically reviewing books for all educational levels. (The journal is now Science Books & Films, published in newsletter format).
1966	Membership passed the 100,000 level. A third annual AAAS Westinghouse Science Writing Award was started, to recognize excellent science writing in newspapers of small circulation.
1967	The Commission on Science Education completed work on Science A Process Approach materials for science in kindergarten and the first and second grades.
1969	The annual meeting in Boston had the largest registration of any in AAAS history (7900). An International Conference on Arid Lands in a Changing World was held in Tucson, Arizona. The Board of Directors appointed the Herbicide Assessment Commission. A new Section on Atmospheric and Hydrospheric Sciences (W) was established.
1970	Mina Rees became the first woman voted as president-elect. Sections on Zoological Sciences (F) and Botanical Sciences (G) were merged to form the Section on Biological Sciences (FG) (designated G in 1973).
1972	The Annual Meeting was held in late December for the last time.
1973	The seventh constitution and bylaws went into effect, and the first completely popular election in AAAS history was held. A special meeting on "Science and Man in the Americas" was held in Mexico City. Offices of Opportunities in Science and International Science and the Committee on Science, Engineering, and Public Policy and Section X General were established. The Congressional Science and Engineering Fellows Program was initiated.
1974	The name of Section T was changed to "Information, Computing, and Communication." Publication of the AAAS *Bulletin* was terminated; a AAAS News section in *Science* was initiated. The AAAS ABA National Conference of Lawyers and Scientists was established. The Mass Media Science and Engineering Fellows program was started.
1975	The Interciencia Association, of which AAAS was a cofounder, was chartered in Venezuela. The first conference of minority women in science was held. The life membership fee was set at 20 times the annual dues.
1976	The first AAAS annual meeting made fully accessible to the physically handicapped was held in Boston. The Committee on Scientific Freedom and

Responsibility was established, and a Consortium of Affiliates for International Programs was organized. The first annual colloquium on science and public policy (Research and Development in the Federal Budget) was held.

1977 The constitution was amended to add a new objective, "to foster scientific freedom and responsibility." The bylaws were amended to provide that new officers began terms immediately following the annual meeting after their election rather than on 1 January. AAAS cosponsored the Nairobi Seminar on Desertification.

1978 Board members visited the People's Republic of China and worked out an agreement for cooperation between AAAS and the Scientific and Technical Association. The constitution was amended to designate the retiring Section Chairmen as representatives of their Sections on the Council.

1979 *Science 80*, a new magazine aimed at a general audience, was launched to enhance the public's understanding of science and technology. The Committee on Climate was established.

1980 The Board authorized a new Scientific Freedom and Responsibility Award. A program of Science, Engineering, and Diplomacy Fellows was initiated. AAAS cosponsored the New Delhi Global Seminar on the Role of Scientific and Engineering Societies in Development.

1981 The Committee on Science, Arms Control, and National Security was established. The AAAS Westinghouse award program was expanded to radio and television reporting. The Association pledged its resources to a major effort at improving science education in the United States. Average membership of almost 137,000 was an all-time high. Domestic circulation of *Science 81* reached 700,000. Section K was renamed "Social, Economic, and Political Sciences."

1982 The Alaska Division was renamed the Arctic Division. *Science* received the George Polk Award in Journalism. A Coalition for Education in the Sciences, comprised of some 60 affiliated societies, was established. AAAS cosponsored an international conference on the direct effects on plants of rising levels of atmospheric carbon dioxide.

1983 *Science 82*, for the second year running, was voted the National Magazine Award for General Excellence. AAAS and the UN Advisory Committee on Science and Technology for Development convened a panel on the roles of women in development. AAAS cosponsored missions to El Salvador and the Philippines to investigate human rights violations.

1984 The Caribbean Division of the AAAS was established. A program on Population, Resources, and the Environment was started. Publication of *Scientific Freedom and National Security*, a quarterly bulletin, began (title and frequency have since varied). An AAAS delegation of forensic scientists visited Argentina to investigate the problem of disappeared persons.

1985 The Board established the AAAS Philip Hauge Abelson Prize and launched Project 2061. A Board delegation to the People's Republic of China was received by Premier Zhao Ziyang. AAAS sent forensic scientists to Argentina to conduct training on identifying human skeletal remains of the "disappeared."

118 / HISTORY

1987 The Annual Meeting was changed from May to February. The constitution was amended to provide appointment of the Section Secretary by the Section Steering Group. A new AAASWestinghouse Award for Public Understanding of Science and Technology was initiated.

1988 A delegation from the Soviet Academy of Sciences participated in the Annual Meeting. A member newsletter, *The AAAS Observer*, was published until December 1989. Project 2061 completed its report, "Science for All Americans" (available from Oxford University Press).

1989 A Public Science Day was inaugurated at the Annual Meeting. AAAS co-sponsored a meeting on the human genome project. *Science* published the discovery of the cystic fibrosis gene. A project began on science for Girl Scouts.

1990 AAAS programs were merged into three Directorates: Science and Policy, International Science, and Education and Human Resources. Project 2061 began the development of a model science curriculum. A science program for Hispanic school children was initiated in Chicago.

1991 President George Bush delivered the keynote address at the Annual Meeting. Science ad sales were brought in-house. The Sub-Saharan Africa program completed a project on cross-sectoral approaches to malaria. President-elect Leon Lederman issued a report on the crisis in science funding.

1992 AAAS launched the *Online Journal of Current Clinical Trials*, the first scientific peer-reviewed journal to be published electronically with print-quality graphics. A new meeting, Science Innovation, was started to highlight tools and techniques in science. International and Science Policy programs provided advice and coordination for scientists in the former Soviet Union, including a new newsletter, which serves as a clearinghouse in this area. *Science* redesigned the magazine and published issues on women in science, minorities in science, and science in Japan, and increased its coverage of science in Europe. In addition, pages devoted to the physical sciences were expanded in the latter half of 1992. AAAS received the first Human Rights Award from the American Psychiatric Association.

RESOLUTIONS AND POLICY STATEMENTS

For the full text of some policy statements see pages 130–134. The following is a selected list of other resolutions and policy statements, 1946–1993, adopted by the Council (and/or Board of Directors and Committees if so indicated).

1946 Appoint an Inter-Society Committee for a National Science Foundation.

1947 Establish the AAAS Committee on Civil Liberties for Scientists. Endorse reconstruction of European educational research facilities.

1948 Protest against the treatment of E.U. Condon by a subcommittee of House of Representatives' Committee on Un-American Activity (Board). Affirm the importance of animals for experimental purposes (Board).

1949 Reject proposed amendments to the Atomic Energy Act (Board). Support the study of the methodology of human ecology by the UN and UNESCO.

1950	Use scientific personnel during the present emergency in a Universal National Service.
1951	Adopt the Arden House Statement on purposes of the AAAS (Board). Support modifications of the McCarran Act for free interchange of scientific knowledge.
1953	Support international cooperation to achieve beneficial utilization of nuclear energy.
1954	Support a uniform policy in awarding government grants for unclassified research. Endorse fluoridation of community water supplies.
1955	Endorse an amendment to McCarran-Walter Act for removal of travel restrictions on foreign scholars. Support government funding for travel to international meetings. Establish an interim committee on the social aspects of science.
1956	Require AAAS Annual Meetings to be held in locations free from racial segregation (Council, by mail ballot). Support placement assistance for refugee Hungarian scientific and technical personnel.
1957	Appoint a AAAS committee on adoption of the metric system. Support revision of the tax laws to encourage gifts to education and science. Support subject matter training for secondary school teachers of mathematics and science. Support increased funding for all areas of basic science research.
1958	Appoint new committees on cooperation among scientists, science in the promotion of human welfare, and public understanding of science (Board, October; Council, December). Establish a committee on council activities and organization. Support AAAS participation in international scientific programs. Support simplified passport regulations to increase international travel and communication. Support the Geneva Conference negotiations on control of nuclear weapons tests. Support federal aid to education (National Defense Education Act).
1959	Support legislation on the metric system. Prepare books and pamphlets to improve the public understanding of science. Appoint a AAAS study committee on international communication. Eliminate the disclaimer affidavit on loyalty required by the National Defense Education Act.
1961	Investigate the feasibility of an international science register and presentation of symposia on foreign science at AAAS Annual Meetings.
1965	Call for settlement of the war in Vietnam. Expression of concern over effects of decreased federal support for research in science and health; call for establishment of AAAS committee to investigate and report.
1966	Express concern and establish a AAAS study committee on use of biological and chemical warfare agents.
1967	Conserve land and water resources. Oppose the Red River (Kentucky) dam.
1968	Call for a field study of wartime use of herbicides (Board and Council).
1969	Set Association membership goals for coming decade (Board). Call for cessation of herbicide use in Vietnam; call for field study of effects of their military herbicide use (repetition of resolution). Extend scientist exchange program between East and West.

1970	Urge national commitment to the metric system. Reaffirm AAAS commitment to equal opportunity.
1971	Endorse metric education. Condemn alleged mistreatment of academics in Mexico. Establish a AAAS Office for women's equality. Prohibit smoking at AAAS meetings.
1972	Support population research. Support assessment of the ecological consequences of the Vietnam war (Senate Bill S-3084). Oppose inclusion of creationism in science curricula (three separate resolutions by the Council, Board, and Commission on Science Education). Conserve nonhuman primates.
1973	Oppose incarceration of foreign scientists in mental hospitals.
1974	Endorse equal opportunity in the sciences and engineering (Board). Improve forest management. Urge safe exit and U.S. assistance to Chilean scholars. Support eligibility for grants and fellowships despite age or lack of institutional connections. Protest violation of scientific norms in the USSR; urge freedom of emigration of Soviet scientists.
1975	Support revision of the copyright law (Board). Protest exclusion of Israel from UNESCO. Urge observance of principles of international cooperation at the Law of the Sea Conference. Oppose discrimination against sexual minorities. Support reinstatement of NSF postdoctoral fellowship program. Establish editorial policy of *Science* magazine (Board; endorsed by Council, February 21, 1976).
1976	Condemn textual deletions in *Science* in the USSR (Committee on Council Affairs). Reduce nuclear weapons levels. Call for reduction of nuclear weapons levels. Call for the respect for human rights in USSR.
1977	Reexamine criteria and processes for selection of AAAS Fellows. Call for issuance of regulations under Rehabilitation Act of 1973.
1978	Require future AAAS annual meetings to be held in states that have ratified the ERA (Board).
1979	Urge retention science education support by the National Science Foundation. Support inclusion of behavioral and social sciences in fields eligible for National Medal of Science. Oppose violation of human rights of Argentine scientists.
1980	Favor conservation of germplasm resources. Deplore genocide in Cambodia. Support integrity of Amazonian ecosystem.
1981	Oppose weapons of mass destruction. Encourage science and engineering education (Board and Council). Urge reestablishment of the President's Science Advisory Committee (PSAC) (Board and Council). Endorse the science policy statement of the Arctic Division. Protect Yanoama people of Brazil.
1982	Oppose forced teaching of creationist beliefs in public school science education (Board and Council). Deny scientific basis for statutory finding on beginning of human life. Support the Endangered Species Act. Endorse immediate ratification of the Genocide Treaty. Support resolution on science education. Support for concurrent Resolution 44 on thermonuclear war. Oppose government restrictions on availability of unclassified research.
1983	Support negotiations on international security, nuclear war, and nuclear weap-

ons. Maintain and improve federal statistical programs. Support taxonomic research and floral and faunal surveys. Statement on federal funding of science (Board).

1984 Urge protection of fundamental rights of scientists. Statement on openness and science and technology. Statement on openness and national security. Appeal on behalf of Academician Sakharov and Dr. Bonner.

1985 Deplore violations of human rights in Nicaragua. Reaffirm scientific peer review of grant proposals.

1986 Statement on Gramm-Rudman-Hollings Emergency Deficit Reduction Act of 1985. Urge reestablishment of the Ethics Advisory Board, U.S. Department of Health and Human Services.

1987 Urge release of all Soviet scientists confined because of beliefs. Commendations for those who have worked to strengthen science education. Endorse the Carnegie report on teacher training and standards.

1988 Commendation to Reagan and Gorbachev for leadership in reducing deployment of nuclear weapons.

1989 Support U.S. return to UNESCO.

1990 Support responsible use of animals in research, testing, and education (Board and Council). Support federal legislation providing immunity for investigations and reporting of scientific fraud (Board).

1991 Support additional funding for contraceptive research. Call for institutional mechanisms and coordinated efforts in environmental research and training. Support availability of RU486 for further research and use as medically indicated. Support a national center for biodiversity. Expressed concern regarding proposed federal ethics rules limiting participation by federal employees in professional societies (Board). Supported U.S. ratification of the International Covenant on Civil and Political Rights.

1992 Changed scheduling to allow election of Fellows by Council prior to annual meeting. Established, on a provisional basis, a new section on linguistics and language sciences. Changed official date of Committee of Council Affairs meeting at annual meeting. Approved change to Pacific Division's bylaws which expanded its territorial definition. Resolution increasing page allotment for SCIENCE in effort to encourage non-biological sciences coverage in reports section of magazine (Board).

1993 Approved changes to SWARM division bylaws which expanded territory and also changed the procedures by which SWARM division officers are elected. Approved formation of a task force to review current section structure.

MEETINGS AND PRESIDENTS

The AAAS president serves three years in the presidential sequence: as president-elect the first year, as president the second year, and as retiring president and chairman of the Board the third year. A list of the presidents of AAAS since its founding in 1848, together with their disciplines and the meetings at which they officiated, follows.

1848 Sept.	Philadelphia	William B. Rogers *(geology)* [acting until the installation of first elected President Redfield] William C. Redfield *(geology)*
1849 Aug.	Cambridge, MA	Joseph Henry *(physics)*
1850 Mar.	Charleston, SC	A. D. Bache *(oceanography)*
1850 Aug.	New Haven	A. D. Bache *(oceanography)*
1851 May	Cincinnati	Louis Agassiz *(glaciology, zoology)*
1851 Aug.	Albany, NY	Louis Agassiz *(glaciology, zoology)*
1852	no meeting	Benjamin Peirce *(physics)*
1853 July	Cleveland	Benjamin Peirce *(physics)*
1854 April	Washington, DC	James D. Dana *(geology)*
1855 Aug.	Providence	John Torrey *(botany)*
1856 Aug.	Albany, NY	James Hall *(geology)*
1857 Aug.	Montreal	J. W. Bailey *(chemistry)* Alexis Caswell *(astronomy)* [successor to J. W. Bailey, deceased]
1858 April	Baltimore	Jeffries Wyman *(medicine)*
1859 Aug.	Springfield, MA	Stephen Alexander *(astronomy)*
1860 Aug.	Newport, RI	Isaac Lea *(geology)*
(no meetings 1861–1865 and no presidents)		
1866 Aug.	Buffalo	F. A. P. Barnard *(astronomy)*
1867 Aug.	Burlington, VT	J. S. Newberry *(geology)*
1868 Aug.	Chicago	Benjamin A. Gould *(astronomy)*
1869 Aug.	Salem, MA	J. W. Foster *(geography)*
1870 Aug.	Troy, NY	William Chauvenet *(astronomy)* T. Sterry Hunt *(geology)* [successor to W. Chauvenet, deceased]
1871 Aug.	Indianapolis	Asa Gray *(botany)*
1872 Aug.	Dubuque	J. Lawrence Smith *(chemistry)*
1873 Aug.	Portland, ME	Joseph Lovering *(physics)*
1874 Aug.	Hartford	John L. LeConte *(entomology)*
1875 Aug.	Detroit	Julius L. Hilgard *(geography)*
1876 Aug.	Buffalo	William B. Rogers *(geology)*
1877 Aug.	Nashville	Simon Newcomb *(astronomy)*
1878 Aug.	St. Louis	O. C. Marsh *(paleontology)*
1879 Aug.	Saratoga Springs	George F. Barker *(chemistry)*
1880 Aug.	Boston	Lewis H. Morgan *(anthropology)*
1881 Aug.	Cincinnati	George J. Brush *(geology)*
1882 Aug.	Montreal	J. W. Dawson *(geology)*
1883 Aug.	Minneapolis	Charles A. Young *(astronomy)*
1884 Sept.	Philadelphia	J. P. Lesley *(geology)*
1885 Aug.	Ann Arbor	H. A. Newton *(mathematics)*

1886 Aug.	Buffalo	Edward S. Morse *(zoology)*
1887 Aug.	New York	S. P. Langley *(physics)*
1888 Aug.	Cleveland	J. W. Powell *(geology)*
1889 Aug.	Toronto	T. C. Mendenhall *(physics)*
1890 Aug.	Indianapolis	George L. Goodale *(botany)*
1891 Aug.	Washington, DC	Albert B. Prescott *(chemistry)*
1892 Aug.	Rochester, NY	Joseph LeConte *(geology)*
1893 Aug.	Madison, WI	William Harkness *(astronomy)*
1894 Aug.	Brooklyn	Daniel G. Brinton *(anthropology)*
1895 Aug.	Springfield, MA	Edward W. Morley *(chemistry)*
1896 Aug.	Buffalo	Edward D. Cope *(paleontology)* Theodore Gill *(zoology)* [successor to Edward D. Cope, deceased]
1897 Aug.	Detroit	Wolcott Gibbs *(chemistry)* [W. J. McGee presided in Gibbs's absence]
1898 Aug.	Boston	F. W. Putnam *(anthropology)*
1899 Aug.	Columbus, OH	Edward Orton *(geology)* Marcus Benjamin *(social sciences)* and Grove Karl Gilbert (geology) [successors to Edward Orton, deceased]
1900 June	New York	R. S. Woodward *(mathematics)*
1901 Aug.	Denver	Charles S. Minot *(medicine)*
1902 June	Pittsburgh, PA	Asaph Hall *(astronomy)*
1902 Dec.	Washington, DC	Ira Remsen *(chemistry)*
1903 Dec.	St. Louis	Carroll D. Wright *(economics)*
1904 Dec.	Philadelphia	W. G. Farlow *(botany)*
1905 Dec.	New Orleans	C. M. Woodward *(mathematics)*
1906 June	Ithaca	William H. Welch *(medicine)*
1906 Dec.	New York	William H. Welch *(medicine)*
1907 Dec.	Chicago	E. L. Nichols *(physics)*
1908 June	Hanover, NH	Thomas C. Chamberlin *(geology)*
1908 Dec.	Baltimore	Thomas C. Chamberlin *(geology)*
1909 Dec.	Boston	David Starr Jordan *(biology)*
1910 Dec.	Minneapolis	A. A. Michelson *(physics)*
1911 Dec.	Washington, DC	Charles E. Bessey *(botany)*
1912 Dec.	Cleveland	E. C. Pickering *(astronomy)*
1913 Dec.	Atlanta	Edmund B. Wilson *(zoology)*
1914 Dec.	Philadelphia	Charles W. Eliot *(education)*
1915 Aug.	San Francisco	W. W. Campbell *(astronomy)*
1915 Dec.	Columbus, OH	W. W. Campbell *(astronomy)*
1916 Dec.	New York	Charles R. Van Hise *(geology)*
1917 Dec.	Pittsburgh, PA	Theodore W. Richards *(chemistry)*
1918 Dec.	Baltimore	John Merle Coulter *(botany)*

1919 Dec.	St. Louis	Simon Flexner *(medicine)*
1920 Dec.	Chicago	Leland O. Howard *(entomology)*
1921 Dec.	Toronto	Eliakim H. Moore *(mathematics)*
1922 June	Salt Lake City	J. Playfair McMurrich *(anatomy)*
1922 Dec.	Boston	J. Playfair McMurrich *(anatomy)*
1923 Sept.	Los Angeles	Charles D. Walcott *(paleontology)*
1923 Dec.	Cincinnati	Charles D. Walcott *(paleontology)*
1924 Dec.	Washington, DC	J. McKeen Cattell *(psychology)*
1925 June	Boulder, CO	Michael I. Pupin *(engineering)*
1925 June	Portland, OR	Michael I. Pupin *(engineering)*
1925 Dec.	Kansas City, MO	Michael I. Pupin *(engineering)*
1926 Dec.	Philadelphia	Liberty Hyde Bailey *(horticulture)*
1927 Dec.	Nashville	Arthur A. Noyes *(chemistry)*
1928 Dec.	New York	Henry F. Osborn *(paleontology)*
1929 Dec.	Des Moines	Robert A. Millikan *(physics)*
1930 Dec.	Cleveland	Thomas H. Morgan *(genetics)*
1931 June	Pasadena	Franz Boas *(anthropology)*
1931 Dec.	New Orleans	Franz Boas *(anthropology)*
1932 June	Syracuse	John Jacob Abel *(pharmacology)*
1932 Dec.	Atlantic City	John Jacob Abel *(pharmacology)*
1933 June	Chicago	Henry N. Russell *(astronomy)*
1933 Dec.	Boston	Henry N. Russell *(astronomy)*
1934 June	Berkeley	Edward L. Thorndike *(psychology)*
1934 Dec.	Pittsburgh, PA	Edward L. Thorndike *(psychology)*
1935 June	Minneapolis	Karl T. Compton *(physics)*
1935 Dec.	St. Louis	Karl T. Compton *(physics)*
1936 June	Rochester, NY	Edwin G. Conklin *(biology)*
1936 Dec.	Atlantic City	Edwin G. Conklin *(biology)*
1937 June	Denver	George D. Birkhoff *(mathematics)*
1937 Dec.	Indianapolis	George D. Birkhoff *(mathematics)*
1938 June	Ottawa	Wesley C. Mitchell *(economics)*
1938 Dec.	Richmond, VA	Wesley C. Mitchell *(economics)*
1939 June	Milwaukee	Walter B. Cannon *(physiology)*
1939 Dec.	Columbus, OH	Walter B. Cannon *(physiology)*
1940 June	Seattle	Albert F. Blakeslee *(genetics)*
1940 Dec.	Philadelphia	Albert F. Blakeslee *(genetics)*
1941 June	Durham, NH	Irving Langmuir *(chemistry)*
1941 Sept.	Chicago	Irving Langmuir *(chemistry)*
1941 Dec.	Dallas	Irving Langmuir *(chemistry)*
1942	no meeting	Arthur H. Compton *(physics)*
1943	no meeting	Isaiah Bowman *(geography)*
1944 Sept.	Cleveland	Anton J. Carlson *(physiology)*
1946 Mar.*	St. Louis	C. F. Kettering *(engineering)*

*(1945 meeting postponed)

1946 Dec.	Boston	James B. Conant *(chemistry)*
1947 Dec.	Chicago	Harlow Shapley *(astronomy)*
1948 Sept.	Washington, DC	Edmund W. Sinnott *(botany)*
1949 Dec.	New York	E. C. Stakman *(plant pathology)*
1950 Dec.	Cleveland	Roger Adams *(chemistry)*
1951 Dec.	Philadelphia	Kirtley F. Mather *(geology)*
1952 Dec.	St. Louis	Detlev W. Bronk *(physiology)*
1953 Dec.	Boston	Edward U. Condon *(physics)*
1954 Dec.	Berkeley	Warren Weaver *(mathematics)*
1955 Dec.	Atlanta	George W. Beadle *(genetics)*
1956 Dec.	New York	Paul B. Sears *(plant ecology)*
1957 Dec.	Indianapolis	Laurence H. Snyder *(genetics)*
1958 Dec.	Washington, DC	Wallace R. Brode *(chemistry)*
1959 Dec.	Chicago	Paul E. Klopsteg *(physics)*
1960 Dec.	New York	Chauncey D. Leake *(pharmacology, history of science)*
1961 Dec.	Denver	Thomas Park *(animal ecology)*
1962 Dec.	Philadelphia	Paul M. Gross *(chemistry)*
1963 Dec.	Cleveland	Alan T. Waterman *(physics)*
1964 Dec.	Montreal	Laurence M. Gould *(geology)*
1965 Dec.	Berkeley	Henry Eyring *(chemistry)*
1966 Dec.	Washington, DC	Alfred S. Romer *(paleontology)*
1967 Dec.	New York	Don K. Price *(political science)*
1968 Dec.	Dallas	Walter Orr Roberts *(astronomy, meteorology)*
1969 Dec.	Boston	H. Bentley Glass *(genetics)*
1970 Dec.	Chicago	Athelstan Spilhaus *(meteorology, oceanography)*
1971 Dec.	Philadelphia	Mina Rees *(mathematics)*
1972 Dec.	Washington, DC	Glenn T. Seaborg *(chemistry)*
1973 June-July	Mexico City (special meeting)	Leonard M. Rieser *(physics)*
1974 Feb.-Mar.	San Francisco	Roger Revelle (oceanography, geophysics, human population)
1975 Jan.	New York	Margaret Mead *(anthropology)*
1976 Feb.	Boston	William D. McElroy *(biology, biochemistry)*
1977 Feb.	Denver	Emilio Q. Daddario *(law, public service)*
1978 Feb.	Washington, DC	Emilio Q. Daddario *(law, public service)***
1979 Jan.	Houston	Edward E. David, Jr. *(physics, electrical engineering)*

** Mr. Daddario served as president from 1 January 1977 through 17 February 1978. Two meetings were held during his presidency.

1980 Jan.	San Francisco	Kenneth E. Boulding *(economics)*
1981 Jan.	Toronto	Frederick Mosteller *(statistics)*
1982 Jan.	Washington, DC	D. Allan Bromley *(physics)*
1983 May	Detroit	E. Margaret Burbidge *(astronomy)*
1984 May	New York	Anna J. Harrison *(chemistry)*
1985 May	Los Angeles	David A. Hamburg *(health science policy)*
1986 May	Philadelphia	Gerard Piel *(science journalism)*
1987 Feb	Chicago	Lawrence Bogorad *(plant biology)*
1988 Feb.	Boston	Sheila E. Widnall *(aeronautics and astronautics)*
1989 Jan.	San Francisco	Walter E. Massey *(physics)*
1990 Feb.	New Orleans	Richard C. Atkinson *(psychology, cognitive science)*
1991 Feb.	Washington, DC	Donald N. Langenberg *(physics)*
1992 Feb.	Chicago, IL	Leon M. Lederman *(physics)*
1993 Feb.	Boston	F. Sherwood Rowland *(chemistry)*

ADMINISTRATIVE OFFICERS, 1848-1993

Principal Administrative Officers

Permanent Secretaries

Spencer F. Baird	1851-1854
Joseph Lovering	1854-1868
F. W. Putnam	1869
Joseph Lovering	1870-1873
F. W. Putnam	1873-1898
L. O. Howard	1898-1920
Burton E. Livingston	1920-1930
Charles F. Roos	1931-1932
Henry B. Ward	1933-1937
F. R. Moulton	1937-1946

Administrative Secretaries

F. R. Moulton	1946-1948
Howard A. Meyerhoff	1949-1953
Dael Wolfle	1954-1955

Executive Officers

Dael Wolfle	1956-1970
William Bevan	1970-1974

Philip H. Abelson *(acting)*..1974-1975
William D. Carey..1975-1987
Alvin W. Trivelpiece...1987-1988
Philip H. Abelson *(acting)*..1989
Richard S. Nicholson..1989-

Associate Administrative Officers

F. S. Hazard, *Assistant Secretary*.....................................1912–1920
Sam Woodley, *Assistant Secretary*....................................1920-1945
Howard A. Meyerhoff, *Executive Secretary*..........................1945-1946
J. M. Hutzel, *Assistant Administrative Secretary*......................1946-1948
Raymond L. Taylor, *Assistant Administrative Secretary*..............1949-1953
    Associate Administrative Secretary............................1953-1967
John A. Behnke, *Assistant Administrative Secretary*.................1952-1953
    Associate Administrative Secretary...........................1953-1956
William T. Kabisch, *Assistant Executive Officer*....................1967-1970
Richard Trumbull, *Deputy Executive Officer*.......................1970-1974
J. Thomas Ratchford, *Associate Executive Officer*...................1977-1989

Treasurers

Jeffries Wyman...1848
A. L. Elwyn..1849-1870
William S. Vaux..1871-1881
William Lilly...1882-1893
R. S. Woodward..1894-1924
John L. Wirt...1925-1940
Carroll W. Morgan...1941-1944
William E. Wrather...1945-1953
Paul A. Scherer..1954-1962
Paul E. Klopsteg..1963-1969
William T. Golden..1969-

Editors of *Science*

Science was founded in 1880 by Thomas A. Edison. John Michels was the first editor. In 1883, Alexander Graham Bell and Gardner Greene Hubbard purchased the magazine and established the Science Company, which published *Science* from 1883 to 1894. The magazine was then sold to James McKeen Cattell. In 1900, the AAAS entered into an agreement with Dr. Cattell to make *Science* the official journal of the Association, and, in 1945, became its owner and publisher.

J. McKeen Cattell...1895-1944
Josephine Owen Cattell and Jacques Cattell........................1944-1945
Willard L. Valentine...1946-1947
Mildred Atwood *(acting)*..1947-1948

Editorial Board, George Baitsell, *chairman*. 1948-1949
Howard A. Meyerhoff. 1949-1953
H. Bentley Glass *(acting)*. 1953
Duane Roller. 1954
Dael Wolfle *(acting)*. 1955
Graham DuShane. 1956-1962
Philip H. Abelson. 1962-1984
Daniel E. Koshland, Jr.. 1985-

FORMER MEMBERS OF THE BOARD OF DIRECTORS (EXCLUDING PRESIDENTS), 1964-1993

Robert McC. Adams. 1985-1988
Mary Ellen Avery. 1989-1993
Francisco J. Ayala. 1989-1993
Robert W. Berliner. 1983-1987
David Blackwell. 1970-1972
Floyd E. Bloom. 1986-1990
Richard H. Bolt. 1969-1976
Lewis M. Branscomb. 1970-1973
Eloise E. Clark. 1978-1981
Kenneth B. Clark. 1975-1976
Mary E. Clutter. 1986-1990
Joel E. Cohen. 1976
Barry Commoner. 1967-1974
Eugene H. Cota-Robles 1988-1992
Martin M. Cummings. 1977-1980
Ruth M. Davis. 1974-1977
Mildred S. Dresselhaus. 1985-1989
Renée C. Fox. 1977-1980
John W. Gardner. 1964-1965
Joseph G. Gavin, Jr. 1988–1992
John H. Gibbons. 1988-1991
Bernard Gifford. 1977-1978
David R. Goddard. 1964-1967
Nancie L. Gonzalez. 1980-1984
Ward H. Goodenough. 1972-1975
Beatrix A. Hamburg. 1987-1991
Caryl P. Haskins. 1971-1974
Hudson Hoagland. 1966-1969
Gerald Holton. 1967-1970
Mike McCormack. 1976-1979
Daniel P. Moynihan. 1972-1973
Dorothy Nelkin. 1983-1987
Phyllis V. Parkins. 1970-1973
Russell W. Peterson. 1978-1981

John C. Sawhill..1979-1980
John E. Sawyer..1982-1986
Chauncey Starr..1974-1977
H. Burr Steinbach...1964-1969
Kenneth V. Thimann...1968-1971
John A. Wheeler...1965-1968
Linda S. Wilson...1984-1988
Chen Ning Yang...1976-1979
Harriet Zuckerman...1980-1984

Policy Statements

EQUAL OPPORTUNITY IN THE SCIENCES AND ENGINEERING

The American Association for the Advancement of Science is formally committed to the principle of equal opportunity for all persons, without regard to irrelevant considerations of sex, race, creed, color, handicap, national origin, or age. It practices this principle in the selection and promotion of its employees and by opening its membership to all who are interested; by encouraging its nominating committees to nominate women, minority, and handicapped scientists and engineers for elective positions; and by attempting to increase the participation of women, minority, and handicapped scientists and engineers in all of its activities.

The Board of Directors recognizes that complex social, economic, and political forces have combined in the past to discourage women, minority, and handicapped persons from entering the sciences and engineering, and to deny those who do enter, equal access to positions of respect and authority. It is the Board's conviction that if each professional association would take all measures within its power to counteract these historic forces, the cause of truly equal opportunity for everyone in the scientific and engineering professions would be significantly advanced. The Board urges the affiliated organizations to join with the Association in this endeavor.

Adopted by the Board of Directors, AAAS, January 1974; revised 15 October 1977

EDITORIAL POLICY, *SCIENCE* MAGAZINE

The Board of Directors is committed to maintaining *Science* as the foremost American journal for the advancement of science. The journal must, in all respects, continue to communicate with and for the scientific community according to the highest standards of objectivity and professional responsibility. It is the Board's responsibility to select the Editor-in-Chief and to obtain assurance, from time to time, that these objectives and criteria of quality are being met.

The Executive Officer, as publisher and chief operating officer of the Association, will exercise general management responsibility and, in close cooperation with the Editor-in-Chief, will see to the strengthening and improvement of *Science* as a primary activity of the AAAS.

The Editor-in-Chief, on behalf of the Board, and in accordance with policies established by the Council, is responsible for the content and professional quality of *Science*, and will determine the merit, suitability, and presentation of material for the journal, taking into account recommendations of reviewers and referees. The Editor-in-Chief will consult the Board from time to time as to plans and policies for *Science*. At least once each year the Board will review the state of the journal and will receive a comprehensive report from the Editor-in-Chief, including the views of the Editorial Board.

Adopted by the Board of Directors, AAAS, 4 April 1975, Endorsed by the Council, AAAS, 21 February 1976; revised by the Board of Directors, AAAS, 11 June 1993

CRITERIA FOR AFFILIATION WITH AAAS

The objectives of the AAAS are "to further the work of scientists, to facilitate cooperation among them, to foster scientific freedom and responsibility, to improve the effectiveness of science in the promotion of human welfare, to advance education in science, and to increase public understanding and appreciation of the importance and promise of the methods of science in human progress." There are many membership organizations and professional societies which have similar aims or have interest in supporting these objectives. Association with each other can be a mutually useful way of furthering these aims. The AAAS has established affiliation of organizations with AAAS as a means of furthering these common purposes. Criteria for such affiliation are set forth below, but final judgment as to whether an organization sufficiently satisfies these criteria shall rest with the Council.

1. Its aims are clearly directed toward, or consistent with, the objectives of the Association.

2. Its program and record of activities demonstrate interest in or substantial support of research, publications, or teaching in science or the advancement of science.

3. It is committed to the principle of equal opportunity as stated on page 130 of this *Handbook*.

4. It has sufficiently large membership (usually at least 200) and has been in existence for a sufficient time (usually at least five years) to give promise of continued support and worthwhile activity.

Adopted by the Council, AAAS, 21 February 1976; revised 29 May 1986

PROCEDURES FOR AFFILIATION WITH AAAS

The president or other appropriate officer of an organization that wishes to affiliate or reaffiliate with AAAS should send a letter to the AAAS Executive Officer, stating why the organization wishes to be considered for affiliation, and designating the AAAS Sections (up to five) in which the applicant would like to enroll. With the letter should be sent:

1. A statement that gives the year of founding; the total number of members; a list of journals and other publications, indicating frequency of their publication; or other evidence of interest in or substantial support of research or teaching in science or the advancement of science.

2. A brief history of the organization, including its aims.

3. A copy of its current constitution and bylaws.

4. A recent list of the members and their addresses.

5. The names and addresses of current officers and the same information for the preceeding four years.

6. A statement of ways in which the applicant visualizes possibilities of cooperation with the AAAS toward the advancement of science.

7. A statement that the applicant is committed to the principle of equal opportunity as stated on page 130 of this *Handbook*.

8. Fifteen copies of a recent issue of each periodical published by the organization.

No later than 1 October, the complete application should be mailed to the

Executive Officer, AAAS, 1333 H Street, N.W., Washington, DC 20005, for presentation to and consideration by the Committee on Council Affairs at its winter meeting. The Committee's recommendations will be submitted to the Council at the following Annual Meeting.

The Committee on Council Affairs requests that each prospective affiliate send an officer to the meeting at which its application will be considered and to the Council meeting at which final action will be taken. Prospective affiliates will be informed in advance of the exact times and locations of these meetings.

Adopted by the Council, AAAS, 29 May 1986

POLICY CONCERNING COUNCIL MEETING AGENDA

Organizations or individuals who wish to present proposals or resolutions for consideration by the Council at its annual meeting should submit them in writing to the AAAS executive officer at least 60 days in advance of that meeting, for review by the Committee on Council Affairs. The committee asks that the following guidelines be observed:

1. All proposals and resolutions should be consistent with the objectives of the Association and deal with matters appropriate for consideration by the council of a scientific organization. Each resolution should be given a concise title.

2. Resolutions should be written in the traditional format, beginning with one or more "Whereas" statement-of-fact clauses and concluding with a "Therefore be it resolved" paragraph which presents a position that follows logically from the stated premises.

3. Proposals and resolutions that deal with technical matters must be accompanied by substantive supporting data and references.

4. Any proposal involving substantial expenditure of AAAS funds—such as a recommendation for the establishment of a study or investigative committee—should be presented in the form of a study proposal, with budget included, so that the financial implications of positive action are clearly stated.

5. Resolutions adopted by the Council are published in the Proceedings Issue of *Science*. Proponents who wish the AAAS to undertake any wider distribution should submit with their resolutions the names and addresses of individuals, organizations, or publications to which they would like to have copies sent.

The committee schedules an open hearing on the day before the Council meeting at the AAAS Annual Meeting to give interested persons an opportunity to speak for or against proposals or resolutions submitted for possible Council action. Urgent matters that arise after the deadline may be submitted to the executive officer, together with a written explanation of why submission was not made at least 60 days in advance of the Council meeting, for consideration by the Committee on Council Affairs at a meeting following the open hearing. At that meeting, the committee also considers any requests submitted to it by individuals who wish to address the Council on a particular agenda item.

Adopted by the Committee on Council Affairs, AAAS, August 1971; revisions approved by the Council, AAAS, 13 February 1978; revisions approved by the Council, AAAS, 10 February 1992.

SUMMARY OF POLICY, GUIDELINES, AND PROCEDURES FOR COMMUNICATION WITH CONGRESS

Policy

1. *Communication with Congress is important.* AAAS officers, members, organizations and staff represent valuable resources to Congress and congressional organizations. AAAS expertise can be helpful to the congressional decision-making process.

2. *AAAS aims to facilitate communications among the scientific and engineering communities and Congress.* Scientific and technical advice and information — as well as policy analyses — can be made available to the Congress in a manner that is not considered lobbying in the traditional or legal sense.

3. *AAAS does not engage in political activities.* No AAAS funds, equipment, supplies, communications facilities, or physical facilities may be used in any political campaign in support of or in opposition to any candidate for public office (at the federal, state, or local level).

4. *AAAS does not, as a general policy, engage in direct or grass roots lobbying.* To retain its tax-exempt status, AAAS must insure that "no substantial part" of its activities constitutes lobbying activities.

5. *The Executive Officer*, consulting as appropriate with other AAAS officers, *will decide when and if an exception to the general AAAS prohibition against lobbying should be made* in the interest of AAAS and the scientific and engineering communities — or in the public and national interest.

6. *Types of communication are varied in nature.* They are both formal and informal, public and private. The great majority of them can be carried out in ways that do not involve lobbying expenditures. The main types of communication are: (a) oral and written *testimony* before a congressional committee; (b) *conferences and study groups* held by AAAS committees and offices; (c) *correspondence and meetings* with Members of Congress and staff; and (d) *reporting and information seeking* from Congress.

Guidelines for Congressional Communications

1. *Congressional requests are important.* In general, groups and individuals acting on behalf of AAAS will comment on specific legislative and budgetary issues *only upon the request of some congressional committee, subcommittee, or other congressional bodies.* A lobbying exemption does not extend to the request of an individual Member of Congress. Responses should be made to all Members of the requesting unit.

2. *The AAAS service ethic is important.* From an overall point of view, AAAS serves society at large in its role as a 501 (c) (3) nonprofit, charitable organization devoted to the advancement of science.

3. Generally, AAAS should seek to express concerns, educate about consequences, and explore options rather than advocate specific legislative actions. However, in any communication that advocates a position — other than in a requested communication — close attention must be given to the requirements for a full and fair treatment of the issue.

4. Activities and communications aimed at gathering information about legislative activities are not considered lobbying.

5. AAAS members should not be asked by AAAS officers, staff, committees, divisions, or other Association entities to take specific action on legislation. More directly, *AAAS does not do any grass roots lobbying* (e.g., influencing AAAS members to oppose or support a bill).

6. *All advocacy communications from AAAS staff require pre-approval by the Executive Officer.*

Reporting

AAAS staff shall report to their office heads *all* contacts made in their capacity as AAAS staff members with Members of Congress and their staffs. Sometimes this will be after the fact. However, good judgment must be applied in the matter of prior notification — especially if the staff member is involved in a personal meeting at Congress. Exempt from this reporting requirement are AAAS-initiated contacts seeking information from Congress (e.g., *Science* reporters working on a story).

Review and Approval Procedures

Pre-approval by the Executive Officer or by the Board of Directors is required for all communications that are highly visible, that might be interpreted as expressing AAAS policy, or that might be interpreted as lobbying.

Annual Plan

As part of the annual AAAS planning and budgeting cycle, an annual plan for congressional activities will be prepared and maintained. Responsibility for developing, updating, and monitoring the plan and associated congressional activities will reside in the Executive Office.

Adopted by the Board of Directors, AAAS, 14 January 1989

Governance

CONSTITUTION

(Seventh; effective January 1, 1973)

Article I. Name

Section 1. The name of this organization shall be the AMERICAN ASSOCIATION FOR THE ADVANCEMENT OF SCIENCE.

Article II. Objectives

Section 1. The objectives of the American Association for the Advancement of Science are to further the work of scientists, to facilitate cooperation among them, to foster scientific freedom and responsibility, to improve the effectiveness of science in the promotion of human welfare, to advance education in science, and to increase public understanding and appreciation of the importance and promise of the methods of science in human progress.

Article III. Membership and Affiliation

Section 1. Members. Any individual who supports the objectives of the Association and is willing to contribute to the achievement of those objectives is qualified for membership. Members shall be eligible to hold elective positions in the Association and shall have the right:

(a) To vote for the President-Elect, other members of the Board of Directors, Council members, members of the Committee on Nominations, and members of the Nominating Committees of the Electorates; to nominate candidates for those positions by petition.

(b) To vote on the recall of elective members of the Board of Directors.

(c) To vote on amendments to the Constitution and to propose amendments to the Constitution and Bylaws by petition.

Section 2. Fellows. Any Member who is deemed to have made a meritorious contribution to the advancement of science may be elected a Fellow of the Association by the Council.

Section 3. Affiliated Organizations. Organizations that meet the criteria for affiliation which shall have been established by the Council may be elected by the Council as Affiliates of the Association. Affiliates shall have such privileges and obligations as the Board of Directors may determine.

Article IV. Electorates

Section 1. The Association shall be apportioned into no less than five Electorates for the purpose of electing members of the Council. Each Electorate shall have a Nominating Committee. Authority to determine the number and names of the Electorates shall be vested in the Council.

Article V. Sections

Section 1. The Association shall be organized in Sections in accordance with the interests of its Members. Each Section shall have a Section Committee. Authority to determine the Section structure shall be vested in the Council.

Article VI. Officers

Section 1. Elective Officers. The elective officers of the Association shall be the President-Elect, the President, the Chairman of the Board (the retiring President), and eight Directors. In an annual general election, the President-Elect and two Directors shall be elected by the Members from slates of candidates presented by the Committee on Nominations. Such slates shall include any nominations made by petition, as prescribed in the Bylaws. The term of the President-Elect shall be three years; he or she shall serve in the second year as President and in the third year as Chairman of the Board of Directors. The terms of the Directors shall be four years. Elective officers may not serve for more than eight consecutive years on the Board.

Section 2. Administrative Officers. The administrative officers of the Association shall be the Executive Officer, the Treasurer, the Section Secretaries, and such others as the Board of Directors may designate. All administrative officers except the Section Secretaries shall be appointed by the Board for such terms as it may determine. Each Section Secretary shall be appointed by the Section Steering Group for a term of four years.

Article VII. Council

Section 1. Duties. In furtherance of the objectives of the Association, the Council shall establish the general policies governing all programs of the Association. Its powers and responsibilities shall include the following:

(a) To review all programs of the Association, including meetings and publications, and to propose actions to the Board of Directors.

(b) To appoint and to supervise committees and commissions to aid the Council in the discharge of its responsibilities, and to terminate such committees and commissions as appropriate.

(c) To provide for the organization of the Association in Sections.

(d) To provide for the apportionment of the Association into Electorates.

(e) To authorize the establishment of regional and local organizations of the Association and to approve their bylaws and amendments thereto.

(f) To establish the criteria for affiliation and to elect organizations as Affiliates of the Association.

(g) To elect Fellows from among the Members of the Association.

(h) To propose to the Members of the Association the recall of elective members of the Board of Directors.

(i) To adopt resolutions and statements on matters affecting the Association.

(j) To propose amendments to the Constitution and to amend the Bylaws.

Section 2. Membership. The Council shall consist of (a) the members of the Board of Directors, (b) the retiring Section Chairmen, (c) delegates from each

Electorate, elected from among and by the members of the Electorate, (d) at least two delegates from the National Association of Academies of Science, and (e) one delegate from each Regional Division. Delegates shall serve terms of three years; they may serve a maximum of two consecutive terms. The President shall serve as chairman of the Council; the Executive Officer shall serve as secretary.

Article VIII. Board of Directors

Section 1. Duties. In consonance with the general policies established by the Council, the Board of Directors (hereinafter called "the Board") shall conduct the affairs of the Association. Its powers and responsibilities shall include the following:

(a) To have, hold, and administer the property and funds of the Association.

(b) To appoint and to supervise committees to aid the Board in the discharge of its responsibilities, and to terminate such committees as appropriate.

(c) To determine the privileges of, and the dues and fees to be paid by, Members, Fellows, and Affiliates.

(d) To appoint the administrative officers, with the exception of the Section Secretaries.

(e) To adopt and to publish an annual budget for the Association and to arrange for an annual independent audit of its accounts.

(f) To conduct the publication program of the Association.

(g) To determine the time and place of meetings of, or meetings sponsored by, the Association, and to have general responsibility for the programs and arrangements for those meetings.

(h) To authorize public statements on behalf of the Association.

(i) To propose amendments to the Constitution and Bylaws.

(j) To report regularly to the Council on its actions.

Section 2. Membership. The Board shall consist of thirteen members: the Chairman of the Board, the President, the President-Elect, the Treasurer, the eight Directors elected for four-year terms, and the Executive Officer, ex officio, without vote. The Executive Officer shall serve as secretary.

Article IX. Amendments

Section 1. Amendments to the Constitution may be proposed by the Board, by the Committee on Council Affairs, by any member of the Council, or by petition signed by at least two hundred Members of the Association. Proposed amendments shall be submitted to the Committee on Council Affairs, through the Executive Officer, for presentation to and preliminary action by the Council. A duly proposed amendment shall be presented at the first Council meeting held sixty or more days after its submission and shall be published to the Members at least thirty days in advance of that meeting. A proposed amendment shall be submitted to the Members of the Association for mail ballot at the time of the next annual general election following the Council meeting at which it was presented if (a) a majority of the Council so votes or (b) it bears the signatures of one thousand or more Members of the Association. The mailing shall include a statement of the Council's position regarding the proposed amendment. A proposed amendment shall require for its adoption a favorable vote of a majority of all Members or, failing that, of two-thirds of the Members who return ballots. A ratified amendment shall become effective upon its adoption.

Article X. Tax-Exempt Status

Section 1. The Association is nonstock and nonprofit. No part of the net earnings of the Association shall inure to the benefit of, or be distributable to, its Directors, officers, or other private persons, except that the Association shall be authorized or empowered to pay reasonable compensation for services rendered and to make payments and distributions in furtherance of the objectives set forth in Article II of the Constitution. No substantial part of the activities of the Association shall be the carrying on of propaganda, or otherwise attempting to influence legislation, and the Association shall not participate in, or intervene in, any political campaign on behalf of any candidate for public office. Notwithstanding any of the provisions of the Constitution, the Association shall not carry on any other activities not permitted to be carried on (a) by a corporation exempt from Federal income tax under Section 501(c)3 of the Internal Revenue Code of 1954 (or the corresponding provision of any future United States Internal Revenue Law) or (b) by a corporation, contributions to which are deductible under Section 170(c)2 of the Internal Revenue Code of 1954 (or the corresponding provision of any future United States Internal Revenue Law).

Section 2. If in any one year the Association is found to be a private foundation, then and in that event, its income for each such taxable year shall be distributed at such time and in such manner as not to subject the foundation to tax under Section 4942 of the Internal Revenue Code, and the foundation shall not engage in any act of self-dealing [as defined in Section 4941(d) of the Internal Revenue Code], and shall not retain any excess business holdings [as defined in Section 4943(c) of the Internal Revenue Code], and shall not make any investments in such manner as to subject the foundation to tax under Section 4944 of the Internal Revenue Code, and shall not make any taxable expenditures [as defined in Section 4945(d) of the Internal Revenue Code].

Article XI. Dissolution

Section 1. In the event of dissolution or termination of the Association, the Board shall, after the payment of all of the liabilities of the Association, dispose of all of the assets of the Association exclusively for the objectives of the Association, in such manner, or to such organization or organizations organized exclusively for charitable, educational, or scientific purposes as shall at the time qualify as an exempt organization or organizations under Section 501(c)3 of the Internal Revenue Code of 1954 (or the corresponding provision of any future United States Internal Revenue Law) as the Board shall determine. Any of such assets not so disposed of shall be disposed of by the Court of Common Pleas of the County in which the principal office of the Association is then located, exclusively for such purposes or to such organization or organizations as said Court shall determine, which are organized and operated exclusively for such purposes.

BYLAWS

Article I. Membership and Affiliation

Section 1. Members. Members shall receive such publications, shall have such additional privileges, and shall pay such dues and fees as the Board may determine. A Member may be dropped from membership for nonpayment of dues.

Fifty-Year Members (Members who have paid dues for fifty years) and Life Members (Members who have paid a life membership fee of such amount as the Board shall have prescribed) shall be exempt from the payment of dues and shall retain all the rights and privileges of membership.

Section 2. Fellows. A Member whose efforts on behalf of the advancement of science or its applications are scientifically or socially distinguished may, by virtue of such meritorious contribution, be elected a Fellow by the Council. The Executive Officer shall annually present to the Council a slate of nominees for such election. Nominations may be made by the Steering Groups of the Section Committees (see Bylaw Article III, Section 3); by the Executive Officer; and by any three Fellows, provided that at least one of the three is not affiliated with the institution of the nominee.

To be included on the slate, a nomination — whatever its source — must have the approval of a majority of the members of the Steering Group of the Section Committee corresponding to the nominee's primary Electorate. However, if the Steering Group of the Section Committee fails to approve a nomination by three Fellows or by the Executive Officer, the nominator(s) may appeal to the Committee on Council Affairs for review and possible reversal of that action. The number of nominees on the slate may not exceed 0.5 percent of the membership of the Association. The number of persons nominated annually by the Steering Group of a Section Committee may not exceed 0.4 percent of the membership who are enrolled in the corresponding Electorate as their primary Electorate.

Section 3. Affiliates. Each Affiliate shall enroll in from one to five Sections of the Association, subject to the approval of the respective Section Committees, including that of Section Y-General Interest in Science and Engineering, with three exceptions: (i) If an Affiliate applies for enrollment only in Section Y, such enrollment shall be automatic. (ii) If an Affiliate's application for enrollment in one or more Sections other than Section Y is not approved by at least one of those Sections, the Affiliate shall be enrolled automatically in Section Y. (iii) If an Affiliate's application for enrollment in two or more Sections, one of which is Section Y, is not approved by one or more Sections other than Section Y, it shall be enrolled automatically in Section Y. Each Affiliate shall appoint a representative to the Section Committee of each Section in which it is enrolled; such representatives must be Members of the Association.

Failure of the Affiliate to appoint within one year a representative to at least one Section Committee will result in initiation of disaffiliation procedures. In addition, failure of the Affiliate to appoint within one year a representative to other Sections in which it is enrolled will result in automatic cancellation of enrollment in those Sections.

At three-year intervals, the Executive Officer shall communicate with all Affiliates to assure that they wish to continue their affiliation. If an Affiliate replies in the negative or fails to reply in due course, disaffiliation shall be automatic and the term(s) of its Section Committee representative(s) shall expire simultaneously with the lapse of affiliation.

Affiliated organizations that no longer meet the criteria for affiliation may be disaffiliated by the Council in accordance with procedures which shall have been established by the Council. Action on a motion to terminate affiliation may be taken only at a meeting of the Council. Such motion shall require for its adoption an affirmative vote of two-thirds of the members present.

Section 4. National Association of Academies of Science. The National Association of Academies of Science, a group composed of two representatives from each of the affiliated academies of science, shall serve as liaison among the academies and between the academies and the Council. It shall elect its officers for such terms as it shall prescribe and shall hold its meetings and otherwise conduct its affairs as it deems desirable. It shall at appropriate intervals elect two delegates to the Council, each of whom shall serve a term of three years. It shall not re-elect a delegate who is completing two consecutive terms.

Article II. Electorates

Section 1. The electorates shall be subsets of the twenty-three Sections of the Association (see Bylaw Article III), consisting of those Members who are enrolled as voting members of the Sections (see Section 3 of this Article).

Section 2. Prerogatives of Electorates. Each Electorate shall be entitled to elect (a) one or more delegates to the Council (one delegate if the Electorate has 2999 or fewer members, two delegates if it has from 3000 to 5999 members, and so on, thereby adding one delegate for each successive increment of 3000 members); (b) the six members of its own Nominating Committee (see Bylaw Article V, Section 3); (c) the Chairman-Elect of the corresponding Section; and (d) the members-at-large of the corresponding Section Committee. The number of Council delegates per Electorate is based on the number of Members enrolled in each Electorate as their primary Electorate. At five-year intervals, on the basis of the number of Members then enrolled in each Electorate as their primary Electorate, the Council shall reconsider the validity of the formula which determines the number of Council delegates to be elected by the Electorates [see (a) above] and, when necessary to insure equitable representation of the Members, shall change the formula and amend the relevant portion of this Section accordingly.

Section 3. Prerogatives of Members of Electorates. Each Member may enroll in one to three Electorates, may vote in each Electorate in which he or she is enrolled, and shall be eligible for election by those Electorates to any position filled by vote of an Electorate, except that no Member may be nominated for office in more than one Section at a time.

Article III. Sections

Section 1. Authority to establish and terminate Sections shall be vested in the Council. The Sections of the Association are:

Section on Mathematics (A)
Section on Physics (B)
Section on Chemistry (C)
Section on Astronomy (D)
Section on Geology and Geography (E)
Section on Biological Sciences (G)
Section on Anthropology (H)
Section on Psychology (J)
Section on Social, Economic, and Political Sciences (K)
Section on History and Philosophy of Science (L)
Section on Engineering (M)
Section on Medical Sciences (N)
Section on Agriculture (O)
Section on Industrial Science (P)
Section on Education (Q)
Section on Dentistry (R)
Section on Pharmaceutical Sciences (S)
Section on Information, Computing, and Communication (T)
Section on Statistics (U)
Section on Atmospheric and Hydrospheric Sciences (W)
Section on Societal Impacts of Science and Engineering (X)
Section on General Interest in Science and Engineering (Y)
Section on Linguistics and Language Sciences (Z)

Section 2. Section Committees.

(a) Function. The affairs of each Section shall be managed by a Section Committee. Each Section Committee shall promote the work of the Association in its own field and may organize subcommittees for that purpose. Under the general direction of the Section Secretary, and within the context of overall plans for scientific meetings of the Association, each Section Committee may arrange such Section contributions to those meetings as it deems desirable.

(b) *Meetings.* Each Section Committee shall meet at least once annually. The Section Chairman shall preside at meetings of the Section Committee. If the Section Chairman is unavailable at any session, the Section Chairman-Elect shall preside. Three members of a Section Committee shall constitute a quorum. A Section Committee may arrange meetings to be held at places and times other than those of Association meetings, but may not incur financial obligation without prior approval of the Board.

(c) *Membership.* Each Section Committee shall consist of (i) the Section Officers: the retiring Section Chairman, the Section Chairman, the Section Chairman-Elect, and the Section Secretary; (ii) four members-at-large; (iii) one representative of each Affiliate that is enrolled in the Section; and (iv) the Council Delegate(s) of the corresponding Electorate. The Section Chairman-Elect shall be elected annually by the Electorate for a three-year term (the second year as Section Chairman and the third year as retiring Section Chairman) to begin immediately following the Annual Meeting held after the election. The Section Secretary shall be appointed by the Section Steering Group for a four-year term to begin immediately following the Annual Meeting held after the appointment. One member-at-large shall be elected annually by the Electorate for a four-year term to begin immediately following the Annual Meeting held after the election. Each representative of an Affiliate shall be appointed by the Affiliate for a three-year, renewable term.

(d) *Representation on the Council.* The Section Committee shall be represented on the Council by the retiring Section Chairman.

(e) *Vacancies.* In the event of a vacancy in the position of retiring Section Chairman, the Section Chairman shall represent the Section Committee at the next Council meeting. In the event of a vacancy in the position of Section Chairman, Section Chairman-Elect, Section Secretary, or member-at-large, the Steering Group (see Section 3 of this Article) shall appoint a replacement for the remainder of the unexpired term. In the event of a vacancy in the position of representative of an Affiliate, the Affiliate shall appoint a replacement for the remainder of the unexpired term. Vacancies shall be filled, through the appropriate means, within a period of ninety days.

Section 3. Steering Groups of the Section Committees.

(a) *Function.* Each Section Committee shall have a Steering Group to (i) take action on policy matters between meetings of the Section Committee; (ii) annually submit to the Executive Officer for presentation to the Council an approved slate of nominees proposed for election as Fellows (see Bylaw Article I, Section 2); (iii) at four-year intervals, appoint the Section Secretary; (iv) in the event of a vacancy in the position of Section Secretary, Section Chairman, Section Chairman-Elect, or member-at-large, appoint a replacement for the remainder of the unexpired term.

(b) *Membership.* Each Steering Group shall consist of eight members: the retiring Section Chairman, the Section Chairman, the Section Chairman-Elect, the Section Secretary, and the four members-at-large. The Section Chairman shall serve as chairman of the Steering Group; the Section Secretary shall be responsible for the fellowship nomination and review process.

Article IV. Officers

Section 1. Duties

(a) The retiring President shall be a member of the Council and of the Executive Committee, and shall serve as Chairman of the Board.

(b) The President shall be a member of the Board, of the Executive Committee, and of the Committee on Council Affairs, and shall serve as chairman of the Council.

(c) The President-Elect shall be a member of the Board and of the Council, and shall serve as chairman of the Committee on Council Affairs.

(d) The Executive Officer shall be a member of the Board without vote; a member of the Council, of the Committee on Council Affairs, of the Executive Committee, and of the Committee on Investment and Finance; shall serve as secretary of the Board, of the Council, and of the Committee on Council Affairs, and as staff officer of the Committee on Nominations; shall be in charge of the Association's offices and shall manage the affairs of the Association in accordance with procedures prescribed by the Board; shall be custodian of the current operating funds; and shall have the authority to enter into contracts for the Association that have been approved by the Board or that are required for the conduct of the Association's activities specifically provided for in the approved annual budgets.

(e) The Treasurer shall be a member of the Board, of the Council, and of the Committee on Investment and Finance, and shall be responsible for the control and administration of all investment funds; endowment, trust, and gift funds; and such other funds as the Board may designate.

(f) The Section Secretaries shall have general responsibility for the work of their Section Committees, arrangements for sectional contributions to scientific meetings of the Association, the fellowship nomination and review process within the Steering Groups of their Section Committees, and such other duties as may be assigned by the Executive Officer.

(g) The Section Chairmen shall preside at meetings of their Section Committees, serve as chairmen of the Steering Groups of their Section Committees, appoint the chairmen of the Electorate Nominating Committees, and fill vacancies on the Electorate Nominating Committees.

(h) The retiring Section Chairmen shall be members of the Council.

Section 2. Requirement for Election or Appointment. Membership in the Association shall be a requirement for election or appointment (a) to the Board, the Council, the Section Committees, and the Committee on Nominations and (b) as officers of the Regional Divisions and the Local Branches.

Membership in the Electorate shall be a requirement for election as Council delegate of an Electorate, member of an Electorate's Nominating Committee, Section Chairman-Elect, Section Secretary, and member-at-large of a Section Committee.

Article V. Nominations and Elections

Section 1. Committee on Nominations.

(a) *Function.* The Committee on Nominations shall annually present to the Members for election by mail ballot at least two nominations for the position of President-Elect and for each additional position to be filled on the Board. It shall not nominate more than one person who is serving for a fourth consecutive year or longer on the Board. It shall not nominate any person who, if elected, would thereby serve for more than eight consecutive years on the Board.

In addition, the Committee shall annually present to members of the Council for election by mail ballot at least two nominations for each position to be filled on the Committee on Council Affairs by vote of the Council.

(b) *Membership.* The Committee on Nominations shall consist of nine members. Eight shall be elected by the Members of the Association from slates presented by the Committee on Council Affairs, for two-year terms; one shall be a member of the Board, appointed annually by the Board. No member shall serve for more than two consecutive years. The terms of four of the popularly elected members shall expire on the last day of the Annual Meeting. New members shall take office immediately following the Annual Meeting held after their election or appointment. The Committee shall annually select one of its members to serve as chairman. The Executive Officer shall serve as staff officer of the Committee.

(c) *Vacancies.* In the event of a vacancy among the elected members of the Committee on Nominations, the Committee on Council Affairs shall appoint a replacement for the remainder of the unexpired term.

Section 2. Annual General Election Procedures. In an annual general election, slates bearing at least two nominations for each of the following positions shall be presented to the Members of the Association for election by mail ballot:

(a) President-Elect
(b) Members of the Board
(c) Members of the Committee on Nominations.

Nominations for (a) and (b) shall be presented by the Committee on Nominations (see Section 1 of this Article). Nominations for (c) shall be presented by the Committee on Council Affairs [see Bylaw Article VI, Section 1 (a)].

Slates of nominees for positions (a), (b), and (c) shall be published to the Members at least sixty days in advance of the issuance of ballots. Additional names may be placed in nomination for any of these positions by petition of at least one hundred Members submitted to the Executive Officer within forty-five days following such publication. Biographical information concerning the nominees shall be published at or about the time ballots are issued.

Section 3. Nominating Committees of the Electorates.

(a) *Function.* Each Electorate shall have a Nominating Committee which shall at appropriate intervals present to the members of the Electorate for election by mail ballot slates of nominees for the following positions:

(i) Council delegates of the Electorate
(ii) Members of the Electorate's Nominating Committee
(iii) Section Chairman-Elect
(iv) Member-at-large of the Section Committee.

At least two nominations shall be presented for each position to be filled. A Nominating Committee shall not renominate a Council delegate who is completing two consecutive terms or a retiring member of the Nominating Committee.

(b) *Membership.* The Nominating Committee of each Electorate shall consist of six members elected by the Electorate for three-year terms. Retiring members shall not be eligible for immediate re-election. The terms of two members shall expire on the last day of the Annual Meeting. New members shall take office immediately following the Annual Meeting held after their election. The chairman shall be one of the two senior members; he or she shall be appointed by the appropriate Section Chairman.

(c) *Vacancies.* In the event of a vacancy on an Electorate's Nominating Committee, the appropriate Section Chairman shall appoint a replacement for the remainder of the unexpired term.

Section 4. Annual Election Procedures of the Electorates. In an annual election, slates bearing at least two nominations for each of the following positions which are to be filled shall be presented to the members of each Electorate for election by mail ballot:

(a) Council delegates of the Electorate
(b) Members of the Electorate's Nominating Committee
(c) Section Chairman-Elect
(d) Member-at-large of the Section Committee.

Nominations for these positions shall be presented to each Electorate by its Nominating Committee [see Section 3 (a) of this Article].

Slates of nominees for these positions shall be published to the Members at least sixty days in advance of the issuance of ballots. Additional names may be placed in nomination for any of these positions by petition of at least fifty members of an Electorate submitted to the Executive Officer within forty-five days following such publication. Biographical information concerning the nominees shall be published at or about the time ballots are issued.

Section 5. Nomination by Petition. Any petition to place additional names in nomination for any position to be filled through election by the Members or by the Electorates shall be accompanied by a curriculum vitae of the nominee and the nominee's statement of acceptance of nomination.

Article VI. Council

Section 1. Committee on Council Affairs.

(a) *Function.* The Committee on Council Affairs shall serve as the executive committee of the Council. It shall (i) prepare the agenda for meetings of the Council; (ii) receive or initiate, coordinate, and advise the Council on reports of Council committees, resolutions, and proposed actions submitted for consideration by the Council; (iii) review applications for affiliation with the Association and petitions to terminate affiliation, and make recommendations thereon to the Council; (iv) at three-year intervals, review the process for nominating and electing Fellows, and make recommendations thereon to the Council; (v) annually present to the Members of the Association for election by mail ballot at least two nominations for each position to be filled on the Committee on Nominations through election by the Members. In addition, the Committee may (i) recommend to the Council appropriate changes in the Constitution and Bylaws; (ii) establish, charge, and, when appropriate, terminate committees to report to the Council on any aspect of Association policy or program or on other matters affecting the advancement of science; (iii) recommend to the Council that it establish and, when appropriate, terminate such committees.

(b) *Membership.* The Committee on Council Affairs shall consist of eleven members: the President; the President-Elect, who shall serve as chairman; the Executive Officer, who shall serve as secretary; and eight members elected from among and by the Council delegates, from slates presented by the Committee on Nominations, for two-year, renewable terms. The terms of four of the eight elected members shall expire on the last day of the Annual Meeting. New members shall take office immediately following the Annual Meeting held after their election.

(c) *Vacancies.* In the event of a vacancy among the elected members of the Committee on Council Affairs, the Committee shall appoint a replacement for the remainder of the unexpired term.

Section 2. Terms of Council Members. The terms of Council members are stated in Constitution Article VII, Section 2. New members shall take office immediately following the Annual Meeting held after their election.

Section 3. Vacancies on the Council. In the event of a vacancy in the position of a Council delegate, the appropriate nominating committee shall fill the vacancy for the remainder of the unexpired term.

Section 4. Council Meetings. The Council shall meet at least once annually. It may hold special meetings at the call of the President or upon the written request of at least one-fourth of the members of the Council submitted to the Executive Officer. If the President is unavailable at any session, the President-Elect shall preside. If neither is available, the Council members in attendance shall elect a chairman for that session. One-half of the members of the Council shall constitute a quorum.

Section 5. Meeting Procedures. Matters to be included on the agenda for action at any regular or special meeting of the Council shall be submitted in writing to the Executive Officer at least sixty days in advance of the meeting. The Executive Officer shall refer such matters to the Committee on Council Affairs for possible inclusion in the written agenda for the meeting.

A matter not included in the written agenda for the meeting may be taken up by the Council only if:

(a) It was submitted in writing at least sixty days in advance and is brought up for consideration by a member of the Council under the item "new business."

(b) Although not submitted in writing at least sixty days in advance, (i) it is brought up for consideration by a member of the Council under the item "new business," and two-thirds of the members present vote to take it up, or (ii) it was proposed by the Board or by the Committee on Council Affairs. In recognition that urgent matters may arise within the sixty days immediately preceding a meeting of the Council, the Committee on Council Affairs shall meet on the day before the Council Meeting at the Annual Meeting to review such matters submitted in writing to the Executive Officer and, by majority vote of the members present, shall determine whether these matters will be included on the agenda as additional items for consideration by the Council. In submitting such matters to the Executive Officer, their proponents shall explain in writing why submission was not made at least sixty days in advance of the Council meeting.

Any matter taken up by the Council shall be considered adopted if a majority of the members present vote in favor of it, with five exceptions: (i) A resolution shall require an affirmative vote of two-thirds of the members present. (ii) A motion to terminate affiliation shall require an affirmative vote of two-thirds of the members present. (iii) A proposed amendment to the Bylaws shall require an affirmative vote of a majority of the entire Council. (iv) A proposed amendment to the Constitution shall require for its submission to the Members of the Association for mail ballot an affirmative vote of a majority of the entire Council. (v) A motion to recall an elective member of the Board shall require for its submission to the Members of the Association for mail ballot an affirmative vote of three-fourths of the members present.

By ruling of the presiding officer, or on motion of any Council member supported by at least one-third of the Council members present and voting, any matter on the agenda that has not yet come to vote, with the exceptions of a proposed amendment to the Constitution or a motion to recall an elective member of the Board, may be submitted to the entire Council membership for a mail ballot, and shall require for its approval a favorable vote of a majority of the Council members, except that a resolution shall require for its adoption a favorable vote of two-thirds of those members.

Section 6. Interim Procedures. During intervals between Council meetings, members of the Council may be polled by mail on matters of Council business, except as precluded by the Constitution and Bylaws, when so authorized by action of the Council, of the Committee on Council Affairs, or of the Board, or on petition signed by at least one hundred Members of the Association and submitted to the Executive Officer. If such a matter involves external action by officers or other representatives of the Association, a summary of arguments for and against the proposed action, approved by the President, shall be submitted with it.

Article VII. Board of Directors

Section 1. Meetings. The Board shall hold at least four meetings a year. It may convene in additional meetings at the call of the Chairman or upon agreement of a majority of its members. If the Chairman is unavailable at any session, the President or President-Elect shall preside. Seven members of the Board shall constitute a quorum.

Section. 2. Executive Committee.

(a) *Function.* The Executive Committee shall act on behalf of the Board between meetings of the Board. All actions taken by the Committee shall be submitted for review and possible further action at the next following meeting of the Board.

(b) *Membership.* The Executive Committee shall consist of the Chairman of the Board, the President, the Executive Officer, and other members of the Board elected annually by the Board.

Section 3. Terms of Board Members. The terms of Board members are stated in Constitution Article VI, Section 1. New members shall take office immediately following the Annual Meeting held after their election.

Section 4. Vacancies on the Board. Within sixty days of the occurrence of any vacancy in an elective position on the Board, the Board shall fill the vacancy for the remainder of the unexpired term.

Section 5. Recall of Elective Members of the Board. Individual elective members of the Board may be recalled by action initiated by any Council member at any Council meeting. If at least three-fourths of the Council members in attendance at the meeting so vote, a proposal to recall shall be submitted, within thirty days of the meeting, to the Members of the Association for mail ballot. A proposal to recall shall require for its adoption an affirmative vote of a majority of all Members or, failing that, of two-thirds of the Members who return ballots, provided that the number of affirmative votes cast is no less than two-thirds of the number of ballots cast in the election at which the Board member was elected. If the vote is to recall, the recall shall become effective at the close of the balloting.

Article VIII. Financial Administration

Section 1. Control and Administration. The deposit, investment, and disbursement of all funds shall be subject to the direction of the Board. The Executive Officer shall be custodian of the current operating funds. The Treasurer shall be responsible for the control and administration of all investment funds; endowment, trust, and gift funds; and such other funds as the Board may designate.

Section 2. Accounting. All incoming funds shall be received by the Executive Officer, entered in the Association's books, and deposited or invested as shall have been prescribed by the Board. The Executive Officer shall keep proper accounts of all financial transactions of the Association. The accounts of the Association shall be audited annually by a certified public accountant selected by the Board.

Section 3. Budget. The Board shall annually adopt a budget allocating funds of the Association for the purpose of carrying out the objectives of the Association.

Section 4. Contracts. The Executive Officer is empowered and authorized to enter into contracts for the Association that have been approved by the Board or that are required for the conduct of the Association's activities specifically provided for in the approved annual budget.

Section 5. Checks. Checks drawn on the accounts of the Association shall bear the signature of any one of several individuals whom the Board shall have authorized to sign checks on behalf of the Association.

Section 6. Fiscal Year. The fiscal year of the Association shall be from January 1 through December 31.

Section 7. Committee on Investment and Finance.

(a) *Function*. The Committee on Investment and Finance shall make recommendations to the Board on the investment of the Association's funds and on financial questions.

(b) *Investment Portfolio*. The securities of the Association may be bought, sold, or exchanged upon the oral order of the Treasurer or, by his written delegation, the Executive Officer or the chairman or the vice-chairman of the Committee on Investment and Finance. Such an oral order shall be followed promptly by written confirmation signed by the Treasurer or the Executive Officer.

(c) *Membership*. The Committee on Investment and Finance shall consist of the Treasurer, the Executive Officer, and other members appointed by the Board. Each appointed member shall serve a term of three years.

Article IX. Publications

Section 1. The publications of the Association shall be those specified by the Board. They shall be issued in such manner as the Board may direct.

Article X. Scientific Meetings

Section 1. The Association shall hold an Annual Meeting and may hold other scientific meetings at such times and places as the Board shall have determined. The programs and arrangements for the Association's meetings shall be under the general cognizance of the Board.

Article XI. Committees

Section 1. Standing Committees. The standing committees of the Association are:

    (a) Committee on Nominations (see Bylaw Article V).
    (b) Committee on Council Affairs (see Bylaw Article VI).
    (c) Executive Committee (see Bylaw Article VII).
    (d) Committee on Investment and Finance (see Bylaw Article VIII).

Section 2. Other Committees.

(a) Committees may be established, charged, and, when appropriate, terminated by the Council, by the Board, and by the Committee on Council Affairs, as provided in Articles VII and VIII of the Constitution and in Article VI of the Bylaws, respectively. In its charge to a committee, the appointing body shall make explicit the term of the committee's effective life. The appointing body may subsequently extend that term if, in its judgment, such extension is desirable. The appointing bodies shall annually review the activities of their committees and, where appropriate, shall provide for rotation of committee membership.

(b) Nominating Committees of the Electorates are described in Bylaw Article V, Section 3.

(c) Section Committees are described in Bylaw Article III, Section 2.

(d) Steering Groups of the Section Committees are described in Bylaw Article III, Section 3.

Article XII. Regional Divisions and Local Branches

Section 1. Regional Divisions and Local Branches of the Association may be established and terminated by the Council. Each Regional Division and Local Branch shall be organized and operated exclusively to carry out, within its respective territory, the objectives of the Association. The Regional Divisions and Local Branches are:

(a) the Pacific Division, which consists of Members of the Association resident in Alberta, British Columbia, Washington, Oregon, California, Idaho, Nevada, Utah, Montana west of the continental divide, and Hawaii (established in 1915);

(b) the Southwestern and Rocky Mountain Division, which consists of Members of the Association resident in Arizona, Colorado, Kansas, Nebraska, New Mexico, Oklahoma, Texas, Wyoming, Montana east of the continental divide, and Sonora and Chihuahua, Mexico (established in 1920);

(c) the Arctic* Division, which consists of Members of the Association resident in Alaska, Yukon, and Northwest Territories and others who meet such requirements as may be established by the Division and approved by the Council of the Association (established in 1951);

(d) the Caribbean Division, which consists of Members of the Association resident in all the islands and countries in or bordering on the Caribbean Sea, including Mexico's Yucatan Peninsula (established in 1984); and

(e) the Lancaster (Pennsylvania) Branch, which consists of Members of the Association resident in Lancaster and vicinity and others who meet such requirements as may be established by the Branch and approved by the Council of the Association (established in 1934).

Section 2. Government. Each Regional Division and Local Branch shall make bylaws for its own government which shall be subject to the approval of the Council of the Association and shall not be inconsistent with the Constitution and Bylaws of the Association. Such bylaws and amendments thereto shall be submitted to the Council, through the Executive Officer of the Association, for approval. Each Regional Division and Local Branch shall elect its officers for such terms as it shall prescribe and shall

---

* The name of the Division was changed from "Alaska" to "Arctic" in 1982.

hold its meetings and otherwise conduct its affairs as it deems desirable, subject to the relevant provisions of the Bylaws of the Association and to such special provisions as the Council of the Association may establish. Each Regional Division and Local Branch shall annually submit to the Board a financial statement and a written report of its activities in a form prescribed by the Executive Officer.

Section 3. Finances. Each Regional Division and Local Branch may annually submit to the Board its proposed budget for the following year. Such budgets shall be comprehensive statements; they shall include estimates of all anticipated expenses, whatever their nature, and all expected income, whatever its source. The Board shall review such proposed budgets and allocate funds for those budget items, in such amounts, as it deems appropriate.

Article XIII. Participating Organizations

Section 1. Organizations whose activities are planned and directed in close relationship with those of the Association may, upon recommendation by the Board and approval by the Council, be designated Participating Organizations. The Board may review the policies of a Participating Organization at any time and make recommendations thereon to the Participating Organization. The Participating Organization shall be invited to have a representative in attendance at such a review. The Participating Organizations are:

(a) the Gordon Research Conferences (designated in 1955);

(b) the Commission on Professionals in Science and Technology (designated in 1972);

(c) the Illinois Science Lecture Association (designated in 1978).

Article XIV. Official Statements

Section 1. The Association shall not be responsible for statements or opinions advanced by any of its officers, or presented in papers or in discussions at meetings of the Association or its Sections, committees, Regional Divisions, or Local Branches, or printed in its publications, except for those authorized by the Board or by the Council.

Article XV. Parliamentary Authority

Section 1. Robert's rules of order, except when inconsistent with the Constitution and Bylaws of the Association, shall govern the meetings of the Council, Board, Sections, and committees.

Article XVI. Amendments

Section 1. Amendments to the Bylaws may be proposed by the Board, by the Committee on Council Affairs, by any member of the Council, or by petition signed by at least one hundred Members of the Association. Proposed amendments shall be submitted to the Committee on Council Affairs, through the Executive Officer, for presentation to and action by the Council. A proposed amendment, if intended for presentation at a Council meeting, must be submitted at least thirty days in advance of the meeting. The Committee on Council Affairs shall present all duly proposed amendments to the Council, either at a meeting (provided that copies have been distributed to the Council members in advance) or by mail. Whether presented at a meeting or by mail, a proposed amendment shall require for its adoption a favorable vote of a majority of the entire Council. If a proposed amendment fails of adoption at a meeting of the Council, the Committee on Council Affairs may subsequently submit it to the entire Council for mail ballot. Such submission shall include a statement of the Committee on Council Affairs' position regarding the proposed amendment. A ratified amendment shall become effective upon its adoption.

ARTICLES OF INCORPORATION OF THE AMERICAN ASSOCIATION FOR THE ADVANCEMENT OF SCIENCE

Commonwealth of Massachusetts

In the Year One Thousand Nine Hundred and Ninety-three

An Act Further Regulating the Incorporation of the American Association for the Advancement of Science.[*]

Be it enacted by the Senate and House of Representatives, in General Court assembled, and by the authority of the same, as follows:

Section 1. Joseph Henry of Washington, Benjamin Peirce of Cambridge, James D. Dana of New Haven, James Hall of Albany, Alexis Caswell of Providence, Stephen Alexander of Princeton, Isaac Lea of Philadelphia, F. A. P. Barnard of New York, John S. Newberry of Cleveland, B. A. Gould of Cambridge, T. Sterry Hunt of Boston, Asa Gray of Cambridge, J. Lawrence Smith of Louisville, Joseph Lovering of Cambridge, and John LeConte of Philadelphia, their associates, the officers and members of the Association known as the "American Association for the Advancement of Science," and their successors are hereby made a corporation by the name of the "American Association for the Advancement of Science." Said corporation is organized and shall be operated exclusively for charitable, scientific, literary, and educational purposes within the meaning of section 501(c)(3) of the Internal Revenue Code of 1986, as amended, including but not limited to the following purposes: (a) to further the work of scientists, to facilitate cooperation among them, to foster scientific freedom and responsibility, to improve the effectiveness of science in the promotion of human welfare, to advance education in science, and to increase public understanding and appreciation of the importance and promise of the methods of science in human progress; and (b) to engage in joint programs and activities and share facilities, with other tax-exempt, charitable, scientific, literary, and educational organizations with a

---

[*]The original Articles of Incorporation were enacted by the Massachusetts Legislature in 1874. In 1993, by the passage of S1524, Chapter 53, the Legislature modified and updated these Articles so that they read as presented here.

view toward achieving economies of scale, programmatic synergies, and overall enhancement and promotion of the charitable, scientific, literary, and educational purposes of said corporation and such other tax-exempt organizations. Said corporation shall have all the powers and privileges and be subject to the restrictions, duties, and liabilities set forth in the General Laws which now or hereafter may be in force and applicable to such corporation, including without limitation the power to engage in any lawful activity in furtherance of its purposes and each of the powers which may be authorized to a corporation under section six of chapter one hundred and eighty of the General Laws.

Section 2. Said corporation may have and hold by purchase, grant, gift, or otherwise, real estate not exceeding five million, five hundred thousand dollars in value, and personal estate in any amount.**

Section 3. Any two of the corporators above named are hereby authorized to call the first meeting of the said corporation in the month of August next ensuing, by notice thereof by mail, to each member of the said Association.

Section 3A. No part of the assets or net earnings of said corporation, current or accumulated, shall inure to the benefit of or be distributable as dividends or otherwise to the directors, officers, or employees of said corporation or to other private persons, except that said corporation is authorized and empowered to pay reasonable compensation for services actually rendered and to make payments and distributions to further its charitable, scientific, literary, and educational purposes.

No director, officer, employee, member of a committee, person connected with said corporation, or any other private individual shall be entitled to or shall share in the distribution of the corporate assets upon the said corporation's dissolution. Upon dissolution or winding up of said corporation's affairs, whether voluntary or involuntary, all of its assets then remaining in the hands of the board of directors shall, after paying or making provision for payment of all of said corporation's liabilities, be distributed, transferred, conveyed, delivered, and paid over only to educational, scientific, literary, or charitable organizations that are exempt from federal income tax under section 501(c)(3) of the Internal Revenue Code of 1986, as amended, and which are not private foundations within the meaning of section 509(a) of the Internal Revenue Code of 1986, as amended, on whatever terms and conditions and in whatever amounts the board of directors may determine, for use exclusively for educational, scientific, literary, or charitable purposes, except that no distribution shall be made to organizations testing for public safety.

No substantial part of the activities of said corporation shall be the carrying on of propaganda or otherwise attempting to influence legislation, and said corporation shall not participate or intervene in, including the publication or distribution of statements, political campaigns on behalf of or in opposition to any candidate for public office, whether by publishing or distributing statements or otherwise.

Section 4. This act shall take effect upon its passage.

**By Special Meeting of the Massachusetts Legislature, in Chapter 180, enacted October 1, 1971, a blanket modification was established which empowered the corporation to hold real and personal property in an unlimited amount.

House of Representatives, May 17, 1993
Passed to be enacted,
Charles F. Flaherty, *Speaker*

In Senate, May 17, 1993
Passed to be enacted,
William M. Bulger, *President*

26 May, 1993
Approved,
at nine o'clock and 29 minutes A.M.

William F. Weld, *Governor*.

INDEX

A

ABA National Conference of Lawyers and Scientists 17
Academies of science, affiliated with AAAS 81
Academy Research Grants 108
Administrative Officers (1848–1993) 126-127
Affiliated Organizations 77–81
Affiliation
  bylaws 139
  constitution 135
  criteria for 131
  procedures 131–132
Agriculture Section 24–25
  Electorate Nominating Committee 62
Amendments
  bylaws 139–140
  constitution 137–138
American Association for Accreditation of Laboratory Animal Care, representative 22
Annual Meeting 87
  Site Committee 12
Anthropology Section 25–26
  Electoral Nominating Committee 62
Arctic Division
  meetings 89
  officers 69
  organization 75–76
Articles of Incorporation 151–152
Astronomy Section 27
  Electorate Nominating Committee 62–63
Atmospheric and Hydrospheric Sciences Section 27–28
  Electorate Nominating Committee 63
Audiocassettes of meetings 86
Audit Committee 12
Awards 99–108
  committees 14–17

B

Behavioral Science Research Award 14, 101–102
Bell Atlantic/AAAS Institute 90
Biological Sciences Section 29–32
  Electorate Nominating Committee 63
Black Churches Health Connection Project 90
Black Churches Project 90
Black Churches/Black Colleges 90
Board of Directors 5
  bylaws 139
  committees of 12–13
  constitution 135
  Executive Committee of 10
  former members 1964–1993 128–129
Books 86
Bylaws 139–151

C

Caribbean Division
  meetings 89
  officers 69
  organization 76
Chemistry Section 32–34
  Electorate Nominating Committee 63–64
Colloquium on Science and Technology Policy 88
Commission on Professionals in Science and Technology 83
  commissioners 22
Committees
  Board of Directors 12–13
  Board-appointed 17–22
  bylaws 147
Compensation Committee 12
Conferences, topical 88
Congress, policy for communication with 133–134
Congressional requests 133

Congressional Science and Engineering Fellowship Program 99
Consortium of Affiliates for International Programs 96
Constitution 134–138
Council 6–9
  Affairs Committee 10
  bylaws 145–147
  constitution 136–137
  meeting agenda, policy 132

D
Dentistry Section 34
  Electorate Nominating Committee 64
Development Committee 13
Diplomacy fellowships 96, 99
Directorates, activities 90–99
Directories 86
Disabled engineering students and faculty, recruitment and retention 94
Dissolution, constitution 138

E
Education programs 90–95
Education Section 35–37
  Electorate Nominating Committee 64
Elections, bylaws 143–145
Electorate 74–75
  bylaws 140
  constitution 135
  Nominating Committees 62–68
Engineering
  students and faculty with disabilities 94
  Committee on Science, and Public Policy 97
  equal opportunity, policy 130
Engineering fellowships 96
  programs 99
Engineering Section 37–39
  Electorate Nominating Committee 64
Engineers, senior 94
Environmental Education Project 91
Environmental Science and Engineering Fellowship Program 99

Equal opportunity in the sciences and engineering, policy 130
Ethics Group, Professional Society 99
European Projects 97

F
Fellowship Programs 92, 96, 99
  Congressional Science and Engineering Fellowship 99
  Environmental Science and Engineering Fellowship 99
  Mass Media Science and Engineering Fellowship 92
  Overseas Diplomacy Fellowship 99
  Science, Engineering and Diplomacy Fellowship 96
  Sloan Executive Branch Science and Engineering Fellowship 99
Films 86
Financial administration, bylaws 147–148
Foreign Science Lecturer Series 96
Forum for School Science 90
Founding, AAAS 111–113
"Frontiers of Science" 83

G
General Interest in Science and Engineering Section 39–42
  Electorate Nominating Committee 65
Geology and Geography Section 42–43
  Electorate Nominating Committee 65
Girl Scouts, Science, and Mathematics: Linkages for the Future 95
Girls and Science: In Touch With Technology 94–95
Global Change Project 95
Gordon Research Conferences
  Board of Trustees, representative 22
  "Frontiers of Science" 83
Governance 135–138
Grants 99–108

H

Health of the Scientific Enterprise Committee 13
Hilliard Roderick Prize 15, 97, 102–103
Hispanic Outreach Program 91
History and Philosophy of Science Section 44–45
   Electorate Nominating Committee 65
History, AAAS 111–129
Human Genome Conference Series 88
Human resources programs 90–95
Human Rights Program, Science and 98

I

Illinois Science Lecture Association 84
Industrial Science Section 45–46
   Electorate Nominating Committee 66
Information, Computing, and Communication Section 46–47
   Electorate Nominating Committee 66
Interciencia Association, representatives 22
International programs 95–97
International Scientific Cooperation Award 17, 107
International Security, Program on Science and 96–97
Investment and Finance Committee 11

J

Journal Distribution Project 96–97
Journalism, Westinghouse Award 15–17, 105–106

L

Law, Scientific Freedom, Responsibility, and 98–99
Lawyers and Scientists, National Conference of 99
Linguistics and Language Science Section 48
Linkages Project 93
Lobbying 133
Local Branches, bylaws 149–150
Long-Range Planning Committee 13

M

Mass Media Science and Engineering Fellowship Program 92
Math Power 92
Mathematics education programs 92
Mathematics Section 48–49
   Electorate Nominating Committee 66
Medical Sciences Section 49–52
   Electorate Nominating Committee 66–67
Meetings 87–89
   bylaws 146
   presidents (1848–1993) 121–126
Membership 72
   bylaws 139
   constitution 135
   electorates 74–75
Mentor Award 17, 107
Milestones, AAAS 113–118
Museums and community groups 92

N

National Association of Academies of Science 82
National Inventors Hall of Fame, National Selection Committee representative 22
Newcomb Cleveland Prize 14, 100–101
Newsletters 88
Nominations
   bylaws 143–145
   Committee 11–12

O

Officers
   administrative (1848–1993) 126–127
   bylaws 142–143
   constitution 136
Online Journal of Current Clinical Trials 86
Opportunities in Science Committee 18–19
Organizations
   affiliated with AAAS 77–81
   other, representatives to 22

Organizations, participating 83–84
  bylaws 150
Overseas Diplomacy Fellowship Program 99

P

Pacific Division
  meetings 89
  officers 69
  organization 76
Parliamentary authority, bylaws 150
Pharmaceutical Sciences Section 52–53
  Electorate Nominating Committee 67
Philip Hauge Abelson Prize 14, 100
Philosophy of Science *see* History and...
Physics Section 53–54
  Electorate Nominating Committee 67
Policy statements 118–121, 130–134
Political Sciences Section 56–57
  Electorate Nominating Committee 68
Presidents
  meetings (1848–1993) 121–126
Prizes 99–108
  Committee 14
Professional Society Ethics Group 99
Program on Science and International Security 96–97
Project 2061 97
Proyecto Futuro 92
Psychology Section 54–56
  Electorate Nominating Committee 67–68
Public Policy, Committee on Science, Engineering, and 97
Public Science Day 92–93
Public Understanding of Science and Technology 90
  Committee 19–20
Public Understanding of Science and Technology Award 15, 104–105
Publications 85–87
  bylaws 148

R

R&D Budget and Policy Program 98
Regional Divisions
  bylaws 149–150
  meetings 89
  membership 73
  officers 69
  organization 75–77
Reports 86
Representatives to other organizations 22
Resolutions 118–121

S

Science 85
  Editorial Board 23
  editorial policy 130
  editors 127–128
Science and engineering fellowship programs 99
Science and Human Rights Program 98
Science and International Security Committee 20
Science and Policy Programs 97–99
Science and Security Colloquium 89
Science Attaché and Foreign Science Lecturer Series 96
Science Books & Films 86
Science education programs 90–95
Science in Africa 95
Science Innovation 87–88
Science Update 91
Science, Engineering, and Diplomacy Fellowships 99
Science, Engineering, and Public Policy Committee 20–21, 97
Science, Technology, and Government Programs 97–98
Scientific Freedom and Responsibility Award 15, 103–104
Scientific Freedom and Responsibility Committee 21–22
Scientific Freedom, Responsibility, and Law Program 98–99
Scientists and engineers, senior 94

Sections
    bylaws 140–142
    constitution 135
    membership 72–73
    organization 74
        Agriculture 24–25
        Anthropology 25–26
        Astronomy 27
        Atmospheric and Hydrospheric Sciences 27–28
        Biological Sciences 29–32
        Chemistry 32–34
        Dentistry 34
        Education 35–37
        Engineering 37–39
        General Interest in Science and Engineering 39–42
        Geology and Geography 42–43
        History and Philosophy of Science 44–45
        Industrial Science 45–46
        Information, Computing, and Communication 46–47
        Linguistics & Language Science 48
        Mathematics 48–49
        Medical Sciences 49–52
        Pharmaceutical Sciences 52–53
        Physics 53–54
        Psychology 54–56
        Social, Economic, and Political Sciences 56–57
        Societal Impacts of Science and Engineering 58–60
        Statistics 61
Security, international 20, 96–97
Sloan Executive Branch Science and Engineering Fellowship Program 99
Social, Economic, and Political Sciences Section 56–57
    Electorate Nominating Committee 68
Societal Impacts of Science and Engineering Section 58–60
    Electorate Nominating Committee 68
Southwestern and Rocky Mountain Division
    meetings 89
    officers 69
    organization 76–77
Special populations programs 90–94
Staff 2
Statements, official, bylaws 150
Statistics Section 61
    Electorate Nominating Committee 68
Sub-Saharan Africa Program 95–96
Symposia series 86–87

T, V

Tax-exempt status, constitution 138
Technology education programs 94
Treasurers (1848–1993) 127

W

Weaponry, workshops on advanced, in the developing world 97
Western Hemisphere Cooperation Project 96
Westinghouse Science Journalism Awards 15–17, 105–106